SULPHUR-CONTAINING DRUGS AND RELATED ORGANIC COMPOUNDS
Chemistry, Biochemistry and Toxicology
Volume 1: Part A
Metabolism of Sulphur Functional Groups

Ellis Horwood Series in
BIOCHEMICAL PHARMACOLOGY
Series Editor: Dr L. A. DAMANI, King's College London, University of London.

SULPHUR-CONTAINING DRUGS AND RELATED ORGANIC COMPOUNDS
Chemistry, Biochemistry and Toxicology
Volume 1: Metabolism of Sulphur Functional Groups. Parts A and B
Editor: L. A. DAMANI, King's College London, University of London.

SULPHUR-CONTAINING DRUGS AND RELATED ORGANIC COMPOUNDS
Chemistry, Biochemistry and Toxicology
Volume 2: Analytical, Biochemical and Toxicological Aspects of Sulphur Xenobiochemistry.
Parts A and B
Editor: L. A. DAMANI, King's College London, University of London.

SULPHUR-CONTAINING DRUGS AND RELATED ORGANIC COMPOUNDS
Chemistry, Biochemistry and Toxicology
Volume 3: Metabolism and Pharmacokinetics of Sulphur-containing Drugs. Parts A and B
Editor: L. A. DAMANI, King's College London, University of London.

SULPHUR-CONTAINING DRUGS AND RELATED ORGANIC COMPOUNDS
Chemistry, Biochemistry and Toxicology
Volume 1: Part A
Metabolism of Sulphur Functional Groups

Editor:

L. A. DAMANI, B.Pharm., M.Sc., Ph.D., M.R.Pharm.S.
Lecturer in Pharmacy
Chelsea Department of Pharmacy
King's College London, University of London

ELLIS HORWOOD LIMITED
Publishers · Chichester

Halsted Press: a division of
JOHN WILEY & SONS
New York · Chichester · Brisbane · Toronto

First published in 1989 by
ELLIS HORWOOD LIMITED
Market Cross House, Cooper Street,
Chichester, West Sussex, PO19 1EB, England
The publisher's colophon is reproduced from James Gillison's drawing of the ancient Market Cross, Chichester.

Distributors:

Australia and New Zealand:
JACARANDA WILEY LIMITED
GPO Box 859, Brisbane, Queensland 4001, Australia

Canada:
JOHN WILEY & SONS CANADA LIMITED
22 Worcester Road, Rexdale, Ontario, Canada

Europe and Africa:
JOHN WILEY & SONS LIMITED
Baffins Lane, Chichester, West Sussex, England

North and South America and the rest of the world:
Halsted Press: a division of
JOHN WILEY & SONS
605 Third Avenue, New York, NY 10158, USA

South-East Asia
JOHN WILEY & SONS (SEA) PTE LIMITED
37 Jalan Pemimpin # 05–04
Block B, Union Industrial Building, Singapore 2057

Indian Subcontinent
WILEY EASTERN LIMITED
4835/24 Ansari Road
Daryaganj, New Delhi 110002, India

© **1989 L. A. Damani/Ellis Horwood Limited**

British Library Cataloguing in Publication Data
Damani, L. A. (Lyaquatali Abdulrasul), *1949–*
Sulphur-containing drugs and related organic compounds.
Vol. 1
Pt. A: Metabolism of sulphur functional groups
1. Sulphur compounds
I. Title
546.7232
Library of Congress Card No. 88–8446

ISBN 0–7458–0215–X (Ellis Horwood Limited)
ISBN 0–470–21257–8 (Halsted Press)

Typeset in Times by Ellis Horwood Limited
Printed in Great Britain by The Camelot Press, Southampton

Table of contents

SULPHUR-CONTAINING DRUGS AND RELATED ORGANIC COMPOUNDS:
Chemistry, Biochemistry and Toxicology

Preface

Sulphur occurs widely in Nature in the elemental state, as H_2S and SO_2, in various sulphide ores of metals, and in the numerous sulphates; its average abundance in the biosphere has been estimated to be around 600 ppm. Sulphur is essential to the life and growth of all organisms — from microbes to man. Most microorganisms and plants can reduce oxidized forms of inorganic sulphur (e.g. sulphate), incorporating the sulphur into the sulphur amino acids (cysteine and methionine). Mammals in general are incapable of using inorganic sulphur, and their need is met by a supply of sulphur amino acids from plant sources.

Sulphur is a common element in many endogenous materials such as amino acids, enzymic and structural proteins, vitamins, co-enzymes and plant secondary metabolites. The biochemistry of endogenous organosulphur compounds has therefore been the subject of much scientific interest (Young, L. and Maw, G. A., *The Metabolism of Sulphur Compounds*, Methuen, London, 1958; Greenberg, D. M., editor, *Metabolism of Sulphur Compounds*, Academic Press, New York, 1975; Anderson, J. W., *Sulphur in Biology*, Edward Arnold, London, 1978; Jakoby, W. B. and Griffith, O. W., editors, *Sulphur and Sulphur Amino Acids*, in Methods in Enzymology, Volume 143, Academic Press, New York, 1987).

In the last two decades, the emphasis in sulphur biochemistry research has shifted towards exogenous synthetic compounds, since sulphur is a common element in numerous industrial, agricultural and medicinal compounds. In the latter case, almost all pharmacological classes are represented, e.g. H_2-receptor antagonist (cimetidine), gout prophylactic and antiplatelet (sulphinpyrazone), mucolytic (acetylcysteine), antirheumatic (sulindac, penicillamine), anaesthetic (thiopentone), cytotoxic (thioguanine) etc. Such drugs, and other sulphur xenobiotics, represent a spectrum of chemical classes, e.g. thioether (cimetidine — a drug; dimethylsulphide — a beer component), sulphoxide (sulindac — a drug; akenyl cysteine sulphoxides — constituents of onion and garlic), sulphone (dapose — a drug), thiol (thioguanine — a drug; propane thiol — a constituent of onion), thione (thiopentone — a drug;

parathion — an insecticide) etc. In some instances the sulphur functional group may be present as a relatively inconsequential structural feature. In many other cases metabolic reactions at the appropriate sulphur functionalities play an important role in disposition and clearance. These biotransformations may also profoundly alter the pharmacological and toxicological properties of the sulphur xenobiotics. Despite this increased interest in recent years in this area of drug metabolism, i.e. in 'sulphur xenobiochemistry', until now a comprehensive compilation of metabolic and toxicological data on sulphur compounds has not been published.

Sulphur-containing Drugs and Related Organic Compounds: Chemistry, Biochemistry and Toxicology is an attempt at collecting and organizing material on sulphur xenobiochemistry into a 'library' of books, which will serve as a useful reference source on this subject. Authors of individual chapters, who are all active investigators, were asked to ensure that the coverage of material was comprehensive. At the same time it was stressed that the editor did not wish to edit a collection of annotated references. Despite the fact that the three volumes cover vastly different types of topics, at varying levels of existing knowledge, the invited contributors have tried to attain a thorough and critical exposition of their subjects.

The first of these volumes, *Metabolism of Sulphur Functional Groups*, is subdivided into parts A and B which together cover the chemical and biochemical reactivity of organic compounds having different types of sulphur functionalities. Volume 2, *Analytical, Biochemical and Toxicological Aspects of Sulphur Xenobiochemistry* (Parts A and B) covers problems and pitfalls in analysis of sulphur xenobiotics, and describes in detail the chemistry and biochemistry of enzymes that mediate various metabolic reactions at sulphur, with emphasis on how such biotransformations can often affect the pharmacology and toxicology of these compounds. Chapters in Volume 3, *Metabolism and Pharmacokinetics of Sulphur-containing Drugs* (Parts A and B), are in the style of monographs, with references to many different, but structurally related, compounds in each therapeutic class. An attempt is made in these chapters to comment on the relationship between chemical structure and metabolism/pharmacokinetics within each class of compounds.

In a multi-authored work of this nature, overlap is almost inevitable. Although this has been kept to a minimum, in many cases it was felt that repetition was desirable, to present the same material from a different perspective, and, through this duplication, to make the chapters self-contained and more readable. Nonetheless an attempt has been made at extensive cross-referencing, between chapters in the same volume, and between the three volumes. In addition to the individual Part indexes, a combined index for all three volumes appears at the end of Volume 3. It is hoped that this collection of detailed, well-referenced reviews from authors actively involved in studying the metabolism of sulphur xenobiotics will lead to conceptualizations from the vast amount of compound specific data that has accumulated and furthermore to identification of fruitful areas for further research.

June, 1988　　　　　　　　　　　　　　　　　　　　　　　　L. A. Damani

1

Aspects of sulphur chemistry, biochemistry and xenobiochemistry

L. A. Damani
Chelsea Department of Pharmacy, King's College London, Manresa Road, London SW3 6LX, UK.

SUMMARY

1. For the purposes of orientation, this editorial introduction provides a brief outline of the history and chemistry of elemental sulphur, and of some of its organo-sulphur derivatives.
2. The main features of *sulphur biochemistry*, i.e. metabolism of endogenous sulphur compounds, are presented, to highlight the importance of such compounds in a variety of biochemical processes in living systems ranging from microbes to man.
3. A historical introduction to the subject of *sulphur xenobiochemistry*, i.e. metabolism of exogenous sulphur compounds in various biological systems, is provided, with reference to many of the early pioneering studies.
4. Finally, the main metabolic options open to various sulphur functional groups are discussed, with particular emphasis on relating observed biochemical reactions to the chemical reactivity and physicochemical properties of the sulphur functionalities.

1.1 INTRODUCTION

1.1.1 History of elemental sulphur

The name of this fascinating second-row element, sulphur, is derived from the Sanskrit *shulbaari* ('shulba'=copper, 'ari'=envious, hostile, an enemy; hence sulphur or shulbaari=an enemy of copper) (Monier-Williams, 1899), through the Latin *sulphurium* or *sulphurum*. Elemental sulphur has been known and used since antiquity, when it was known as *lapis ardens* or brimstone (burning stone). Man has invested this material with magical properties, in view of the fact that it would often burn during lightning, emitting highly noxious sulphurous fumes. The history of sulphur can be traced to the days of Sodom and Gomorrah; destruction of these twin cities involved in some mysterious way brimstone and fire (The Bible, *King James VI version*, *Genesis* 13, v. 10; 19, v. 24–28). Ancient Greek and Roman writers were familiar with the use of burning sulphur as a means of cleansing and purifying dwellings. Thus Homer in his classic work *The Odyssey* ('The Battle of the Hall', Book 22; translated by E. V. Rieu, Book Club Associates Edition, Guild Publishing, London, 1987, pp. 328–340) had Odysseus use burning sulphur to fumigate the house after the slaughter of one hundred of his wife's suitors. The nature of the chemical combination of sulphur with oxygen was only firmly established three millennia later, around 1800, through the work of Priestly (1790), Scheele (1777) and Lavoisier (1783). It was these studies that led to the recognition that sulphur was not some mystical complex material, but was an element.

1.1.2 History of endogenous or protein sulphur

That sulphur was a common element in proteinaceous material was recognized in the early 18th century, but the nature of the endogenous sulphur was not then known. The disulphide cystine (**1**) was probably the first amino acid to be discovered. Wollaston (1810) isolated 'cystic oxide' from a urinary calculus, but failed to notice that this organic compound contained sulphur. Baudrimont and Malaguti (1837) discovered the presence of sulphur in cystic oxide or cystine, as it was by then called, and Thaulow (1838) proposed the correct empirical formula, $C_6H_{12}N_2O_4S_2$. However, several decades elapsed before its structure was established firmly (Friedmann, 1902, 1903), and its occurrence in proteins was demonstrated (Mörner, 1899). Methionine (**2**) was not discovered until 1922 (Mueller, 1922) and it was only then that a clear understanding of the nutritional significance of cysteine (**3**), cystine and methionine emerged. In the interim period between cystine–cysteine and methionine discoveries, deRey–Pailhade (1888) isolated 'philothion' (glutathione), which was later shown by Hopkins (1929) to be a tripeptide.

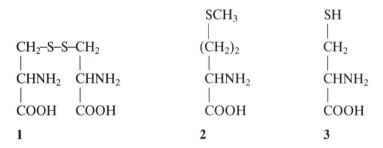

Since these early landmarks in *sulphur biochemistry*, the importance of a wide variety of endogenous sulphur compounds in numerous biochemical reactions, has been recognized. Apart from protein (amino acid) sulphur, this element is present in vitamins (biotin, thiamine), coenzymes (coenzyme A) and numerous plant secondary metabolites. The biochemistry of such endogenous organosulphur compounds has therefore been the subject of much scientific interest (Young and Maw, 1958;

Greenberg, 1975; Jakoby and Griffith, 1987). It is now clear that many of the reactions that these compounds participate in are absolutely reliant on unique aspects of sulphur functional group reactivities.

1.1.3 History of exogenous sulphur
The preparation of carbon disulphide by W. A. Lampadius in 1796 was an important milestone in the history of sulphur chemistry, since it could be used as the starting material for the synthesis of a large number of organosulphur compounds. Synthetic organic chemistry of sulphur really commenced with the discovery of the xanthate reaction by Zeise in 1815, about the time that chemistry was beginning to take shape following the promulgation of the theories of Dalton (see Reid, 1963). The xanthate reaction and its application in the manufacture of rayon and cellophane from cellulose is discussed in Chapter 4 (section 4.3.3) in this volume. Amongst the other major highlights in synthetic organosulphur chemistry were the preparation of ethanethiol (Zeise, 1834; Liebig, 1834), synthesis of dithiocarbamic acids (Debus, 1850) and synthesis of dithizone, or 1,5-diphenylthiocarbazone (Fischer, 1878). Carbon disulphide served as the starting material for many of the early organosulphur componds, e.g. thioureas, dithiocarbamates, dithioic acids and esters, etc.

The history of the metabolism of exogenous or xenobiotic sulphur is linked to an extent to that of endogenous sulphur. It was recognized early on that sulphur from excess dietary proteins was ultimately oxidized to sulphate, which appeared in urine. The realization that cysteine and methionine played a crucial role in nutrition led to the search of other sulphur compounds that could substitute for these essential amino acids in supporting growth in animals fed diets deficient in these components. One of the criteria used in making such an assessment was the ability of such exogenous organosulphur compounds to increase the output of sulphate in urine. Amongst the early compounds studied in this respect were thioglycollic acid, thiourethane and ethanethiol (Smith, 1892, 1893, 1894); each of these, on administration, led to an increase in total urinary sulphate in experimental animals. These early studies on intermediary protein metabolism therefore established that endogenous or exogenous sulphur could be oxidised in mammals to sulphate, which was then eliminated in the urine. Subsequent studies demonstrated that oxidation at the sulphur atom was an important metabolic reaction, affording various intermediate *S*-oxygenated derivatives *in vivo*. The first such example was probably phenothiazine sulphoxide, a metabolite produced *in vivo* from the anthelmintic phenothiazine (Clare, 1947), although there was an earlier unauthenticated report of methylene blue undergoing conversion to its sulphone *in vivo* (Underhill and Classon, 1905). The early history of xenobiotic sulphoxide and sulphone formation is given in Table 1, the data being adapted from the review by Mitchell and Waring (1986). Later studies have demonstrated that sulphur functionalities can participate in a variety of metabolic reactions in addition to oxidations. These are discussed more fully in section 1.5 of this chapter.

1.2 SOURCES OF SULPHUR

1.2.1 World reserves of sulphur
The commercially exploitable forms of sulphur in the biosphere are as follows:

Table 1 — Early history of xenobiotic sulphoxide and sulphone formation (adapted from Mitchell and Waring, 1986)

Reference	Compound (metabolite identified)	Species
Underhill and Classon (1905)	Methylene blue (sulphone?)	Man, dog, cat, rabbit
Clare (1947)	Phenothiazine (sulphoxide)	Pig, sheep, cow
Rose and Spinks (1948)	p-Methylthioaniline (sulphoxide)	Rabbit, rat, mouse
Salzman et al. (1955)	Chlorpromazine (sulphoxide)	Man, dog
March et al. (1955)	'Systox' (sulphoxide and sulphone)	Mouse, cockroach
Snow (1957)	Diethyl disulphide (sulphone)	Rabbit, guinea pig, mouse
Burns et al. (1957)	4-(Phenylthioethyl)-1,2-diphenyl-3,5-pyrazolidinedione (G-25671) (sulphoxide)	Man
Walkenstein and Seifter (1959)	Promazine (sulphoxide)	Dog
Gillette and Kamin (1960)	4,4'-Diaminodiphenylsulphide (sulphoxide)	Guinea pig (in vitro data)
Kane (1962)	Ethionamide (sulphoxide)	Rabbit
Zehnder et al. (1962)	Thioridazine (sulphoxide and sulphone)	Rat

elemental sulphur, hydrogen sulphide occurring in natural gases, sulphur-containing compounds in petroleum crudes, and various metal sulphides (FeS, CuS). However, quantitatively the most important sulphur reserves on earth are gypsum ($CaSO_4$), coal and oil shales (Bixby, 1978). These vast reserves are, however, not directly suitable for recovery of sulphur, although their consumption does contribute to release of sulphur (e.g. as sulphur dioxide) into the environment which is ultimately usable by plants (see below). All living organisms utilize sulphur in one of its many oxidative states. Microorganisms and plants utilize inorganic sulphate from the soil for the production of organic sulphur compounds (cysteine, methionine). Animals in general rely on plant sources for these essential sulphur amino acids; excess dietary sulphur is metabolized to sulphate and returned to the soil in excreta. These biological interconversions of inorganic and organic sulphur in various organisms therefore contribute to the cycling of sulphur in the biosphere. However, net losses of sulphur from the biosphere do occur through removal in feed crops and sediments. Therefore additions of sulphur from the sulphur reserves (coal, oil shales) to the biosphere are important in maintaining sulphur levels. In addition, localized sulphur imbalances within the biosphere (e.g. in crop fields) need correcting via addition of sulphur fertilizers (see Chapter 4, section 4.2).

1.2.2 Mining or recovery of elemental sulphur

Elemental sulphur deposits occur in various parts of the world, e.g. Texas, Alaska (USA), Sicily, Japan, France, Mexico and South America. It has been mined from subterranean deposits in the USA and Mexico by the *Frasch process* (Lundy, 1949); this involves melting the sulphur with superheated water and lifting the molten sulphur with air. Increasing amounts of sulphur have been obtained in the last two decades from hydrogen sulphide in natural gases and from gaseous industrial wastes containing oxides of sulphur. In the *Claus process* (Giusti, 1965), hydrogen sulphide (H_2S) in natural gases is trapped in an alkaline solution and subsequently expelled by heating. A third of the H_2S is converted to sulphur dioxide (SO_2) in a controlled combustion in the presence of sufficient oxygen. The mixture of H_2S and SO_2 is then reacted at high temperatures in a catalytic converter to produce elemental sulphur. Elemental sulphur may also be obtained from pyrites (metal sulphides, FeS, CuS) by heating the material to high temperatures and condensing the released sulphur vapour; however, most of the sulphur in pyrites is obtained by roasting the material to release SO_2, which is processed mainly to sulphuric acid. The total annual production of sulphur is about 60 million tons, very closely matching the present industrial and agricultural requirements (see Chapter 4, section 4.1).

1.3   ASPECTS OF SULPHUR CHEMISTRY

1.3.1   Structures of elemental sulphur

Sulphur (atomic number 16), a yellow crystalline solid with a density of 2, is the second element of group VIB of the periodic table. Sulphur has a mean atomic weight of 32.064 (±0.003) as found in nature, and shows essentially non-metallic covalent chemistry. Its electronic configuration is $1s^22s^22p^63s^23p^4$. All modifications of solid crystalline sulphur contain either (a) rings of sulphur atoms, which may have from 6 to 20 atoms and are referred to as cyclohexasulphur, cyclooctasulphur, etc., *cyclooctasulphur* (S_8) being the most stable of these rings, or (b) chains of sulphur atoms, or catenasulphur. The eight-atom ring form of the molecule, *cycloocta-sulphur* (S_8), is the most common form of sulphur; this exists in three main allotropes (crystal forms). *Orthorhombic sulphur*, S_α, is the most stable allotrope, and occurs as large yellow crystals in volcanic deposits. On slow heating to 95.5°C, this allotrope is converted to *monoclinic sulphur*, S_β, which melts at 119.25°C. The enthalpy of the $S_\alpha \rightarrow S_\beta$ transition is small and the process is slow so that it is possible by rapid heating of S_α to attain the melting point of S_α, 112.8°C (Cotton and Wilkinson, 1972). A further *monoclinic* form, S_γ, melts at 106.8°C and is obtained by slow crystallization of sulphur from ethanolic ammonium polysulphide solutions.

When the three allotropes of *cyclooctasulphur* (S_8) melt, the structural unit in the yellow transparent mobile liquid is still intact, i.e. S_8 rings are still present. However, when this liquid is heated further, the colour darkens and, around 160°C, the rings break and form spiral chains, with a concomitant increase in viscosity. The viscosity reaches a maximum at about 200°C, and thereafter falls until, at the boiling point of 446.6°C, the sulphur is again a rather mobile, dark red liquid. The gaseous phase contains a mixture of S_n species ($n = 2$–8) in a temperature-dependent equilibrium. The sulphur vapour can be recondensed by cooling to afford 'sublimed sulphur', although the process clearly is not one of true sublimation.

1.3.2   Special characteristics of organosulphur chemistry

Sulphur is the element in group VIB of the periodic table placed just below oxygen; it is followed by selenium, tellurium and polonium. Sulphur is a second-row element, positioned two places to the right of carbon. In view of this, the chemistry of organosulphur compounds is best discussed by comparing these compounds with their oxygen- or carbon-substituted analogues. There are many similarities between the first and second elements in a given periodic table column. This is certainly true of group VIB, as is evident from the many similar properties of thiols and alcohols, or sulphides (thioethers) and ethers. However, there are many differences as well, these accounting for many of the unique features of organosulphur chemistry. The similarities between oxygen and sulphur arise from the fact that they have the same number of outer-shell electrons. The differences between them arise from the fact that sulphur possesses an extra inner-electron shell, beneath the outer, bonding shell. Relative to oxygen, sulphur has a larger atomic core (1.02 vs. 0.73 Å), and its outer electrons are further from the nucleus and more shielded from its attractive force. In addition, the outer shell in sulphur contains not only the s and p orbitals, which dominate bonding by seeking the full octet, but also empty d orbitals. The valenay is not limited to two, as in the case of oxygen, since the empty d orbitals may be utilized in bonding by being the recipients of the covalent electrons. Sulphur can therefore expand its octet and form many hypervalent componds (SO_2, SO_3, SF_4, SF_6, suiphoxide, sulphone).

The electronic differences between sulphur and oxygen are responsible for many of the unique aspects of organosulphur chemistry. The lower electronegativity of sulphur compared with oxygen (2.44 vs. 3.5) lessens the ionic character of organosul-phur compounds that are formally analogous to those of oxygen, and decreases considerably the importance of hydrogen bonding in such compounds. The larger size of the sulphur atom results in the non-bonded pairs of electrons' being more polarizable (i.e. 'softer') than those on oxygen. These electron pairs are therefore better nucleophiles, but weaker bases (to H+). Thus, Lewis-base lone-pairs on sulphur centres prefer to complex with soft Lewis acids (e.g. Hg^{2+}) rather than with hard Lewis acids (e.g. H^+). Compared with ethers, ketones and amines (harder Lewis bases) which are very good hydrogen bond acceptors, sulphur centres in thiols for example are poor at hydrogen bonding. Therefore, despite the greater electro-negativity of oxygen compared with sulphur (see above), thiols and carbodithioic acids are stronger acids than the corresponding oxygen analogues. For similar reasons, organosulphur compounds (e.g. thioethers, thiones, disulphides) generally tend to be more lipophilic and diffuse more readily across biological membranes than their related oxygen analogues which interact more favourably with water through hydrogen bonding (Hanzlik, 1984).

Another unique feature of organosulphur chemistry arises from the fact that normal π-orbital bonds are unstable and only rarely occur with sulphur atoms. This is mostly because of the much greater size and longer bond lengths of sulphur compared with those of carbon; the sideways overlap of parallel p orbitals necessary to form a stable π orbital is less extensive, making the π bond very weak. Thus doubly bonded thioketones and thioaldehydes ($R_2C=S$, $RCH=S$) are rare and more reactive than their oxygen analogues. Many thiocarbonyl compounds, especially unsaturated azaheterocycles (e.g. 6-mercaptopurine, methimazole), tend to exist in their eneth-iol tautomeric forms. On the other hand, thione derivatives of carboxylic acids (e.g. thioesters, **4**, thionamides, **5**, thiourea, **6**) are stabilized through conjugation or delocalization. The final unique aspect of sulphur is its ability to form more than four covalent bonds, because of its ability to accept outside electron pairs into its empty d orbitals. This allows sulphur to form several hypervalent compounds (sulphoxides, sulphones).

$$
\begin{array}{ccc}
\overset{\displaystyle S}{\underset{\displaystyle \|}{}} & \overset{\displaystyle S}{\underset{\displaystyle \|}{}} & \overset{\displaystyle S}{\underset{\displaystyle \|}{}} \\
\text{R–C–OR}^1 & \text{R–C–NR}^1\text{R}^2 & \text{H}_2\text{N–C–NH}_2 \\
\mathbf{4} & \mathbf{5} & \mathbf{6}
\end{array}
$$

1.3.3 Oxidation of divalent sulphur

It has long been recognized that divalent sulphur groups, because of the available unshared electron pairs, are nucleophilic and can be oxidized to a variety of covalent compounds differing in oxidation state at the sulphur atom (Table 2). The sulphur atoms are usually less susceptible to steric hindrance than oxygen or carbon centres. The addition of any electron-deficient species to sulphur can be formally regarded as a two-electron oxidation, i.e. oxidation need not always imply oxygen addition. Divalent sulphur compounds (e.g. thioether, thiol) contain sulphur in its lowest oxidation state (-2). The oxidation numbers or states of the oxidation products of these compounds may be computed simply by adding to this lowest state $+2$ for each bond to a more electronegative atom (N, O, halogen) but nothing for bonds to carbon or hydrogen (or unshared pairs) (Hendrickson *et al.*, 1970). Sulphides can be converted to the sulphoxides and sulphones by reacting with peracids; excess oxidant results in complete oxidation to the sulphone, but careful control of the conditions will afford the intermediate sulphoxides. Thiols (oxidation number -2) can be converted via three two-electron oxidation steps to sulphonic acids (oxidation number $+4$) by most strong oxidants (e.g. H_2O_2, $KMnO_4$). The intermediate sulphenic and sulphinic acids are not easily obtained even under carefully controlled conditions. Indeed, sulphenic acids (RSOH) are very unstable and cannot usually be isolated. Sulphinic acids (RSO_2H) on the other hand are moderately stable compounds, and two such compounds, cysteinesulphinate and hypotaurine, are found in biological tissues.

The nature of the sulphur–oxygen bond in sulphoxides and sulphones needs special mention. Two designations are currently used:

$$
\begin{array}{lcll}
\overset{\displaystyle O}{\underset{\displaystyle \|}{}} & & \overset{\displaystyle O^-}{\underset{\displaystyle |}{}} & \\
\text{R–S–R}^1 & \text{or} & \text{R–S}^+\text{–R}^1 & \text{(sulphoxides)}
\end{array}
$$

$$
\begin{array}{lcll}
\overset{\displaystyle O}{\underset{\displaystyle \|}{}} & & \overset{\displaystyle O^-}{\underset{\displaystyle |}{}} & \\
\text{R–S–R}^1 & \text{or} & \text{R–S}^{2+}\text{–R}^1 & \text{(sulphones)} \\
\underset{\displaystyle O}{\overset{\displaystyle \|}{}} & & \underset{\displaystyle O_-}{\overset{\displaystyle |}{}} &
\end{array}
$$

The double-bonded designation focuses on the covalent nature of the S–O bond, whereas S^+–O^- or S→0 indicates a polar bond. The doubly bonded designation is

Table 2 — Oxidation states of sulphur in covalent compounds encountered in sulphur xenobiochemistry

Oxidation state	Example of sulphur functional groups
−2	R–SH (thiols), R–S–R¹ (sulphides), R–S–glycoside (*S*-glycosides), R–S–CO–R¹ (thiol esters), $$\overset{\displaystyle S}{\overset{\|}{R-C-R^1}}\text{(thioketones)},\quad \overset{\displaystyle S}{\overset{\|}{R-C-SH}}\text{(dithioic acids)}$$
−1	R–S–S–R¹ (disulphides), R–S–SO$_3^-$ (thiosulphate esters)
0	$$R-S-OH \text{ (sulphenic acids)},\quad \overset{\displaystyle O}{\overset{\|}{R-S-R^1}}\text{(sulphoxides)},$$ $$\overset{S\to O}{\overset{\|}{R-C-R^1}}\text{(}S\text{-oxides)},\quad \overset{\displaystyle R}{\overset{\|}{\underset{+}{R-S-R^1}}}\text{(sulphonium ions)}$$
+2	$$\overset{\displaystyle O}{\overset{\|}{R-S--OH}}\text{(sulphinic acids)},\quad \overset{\displaystyle O}{\underset{\overset{\|}{O}}{\overset{\|}{R-S-R^1}}}\text{(sulphones)},$$ $$\overset{O\leftarrow S\to O}{\overset{\|}{R-C-R^1}}\text{(}S,S\text{-dioxides)}$$
+4	$$\overset{\displaystyle O}{\underset{\overset{\|}{O}}{\overset{\|}{R-S-OH}}}\text{(sulphonic acids)}$$
+6	$$\overset{\displaystyle O}{\underset{\overset{\|}{O}}{\overset{\|}{R-O-S-OR^1}}}\text{(sulphates)}$$

normally used in the literature, since the S–O bonds are more than a single bond. The S=O bonds in sulphoxides are polar; oxygenation can therefore lead to a considerable change in the chemistry of the parent compound. The nucleophilicity decreases, whereas acidity, polar character, water solubility and leaving-group reactivity increases. The higher oxidation states of sulphur (e.g. sulphones) can act as electrophiles and are often found to be good leaving groups in nucleophilic displacements involving the −SH group of glutathione.

The above brief discourse on organosulphur chemistry is intended to indicate the underlying reasons for the unique behaviour of sulphur centres in chemical and biological reactions, as compared with reactions at oxygen or carbon. Individual chapters in this volume will deal further with the chemistry of the various sulphur functional groups (thioether, sulphoxide, etc.) with a view to interpreting these facts in terms of their biochemical reactivities. More detailed accounts of organosulphur chemistry are available in several classic texts (Kharasch and Meyers, 1966; Oae, 1977; Jones, 1979).

1.4 ASPECTS OF SULPHUR BIOCHEMISTRY

1.4.1 Definition of sulphur biochemistry

Sulphur is one of the major elements essential in nutrition for life and growth of all organisms — microbes, plants and animals. An idea of an organism's requirements for sulphur may be gained from the elemental composition of organisms; sulphur accounts for about 0.2–0.7% of the dry mass of most organisms (Anderson, 1978). The majority of this sulphur is in the form of sulphur amino acids (cysteine, cystine and methionine) in proteins, although a large number of other sulphur-containing endogenous compounds have been isolated and shown to participate in a variety of important biochemical processes. Since the major environmental source of sulphur is inorganic sulphate (SO_4^{2-}, oxidation state +6), whereas that found in organisms is in reduced form (−SH, −S−, oxidation state −2), it is clear that there must exist biochemical mechanisms for sulphur reduction. Most microorganisms and plants can reduce oxidized forms of inorganic sulphur, incorporating the sulphur into the sulphur amino acids. Mammals are incapable of using inorganic sulphur, and their need is met by a supply of sulphur amino acids in their diets. These sulphur compounds then act as the starting point for the synthesis of the vast number of other endogenous organosulphur components (Table 3). The physiological importance of such *primary sulphur metabolites*, i.e. compounds considered as essential to the survival and growth of organisms, is discussed in several books (Young and Maw, 1958; Greenberg, 1975; Anderson, 1978; Jakoby and Griffith, 1987). In microbial and plant systems, various *secondary sulphur metabolites* have also been identified (e.g. penicillins, cephalosporins, glucosinolates, etc.); these compounds are not considered as essential to life.

Sulphur biochemistry may be defined as the study of the metabolism of naturally occurring sulphur compounds in living systems, i.e. a study of their biosynthesis, utilization and ultimate degradation. On the one hand it could include the study of the assimilatory and dissimilatory sulphate reduction in plants for the biosynthesis of sulphur amino acids. On the other hand, it could include the further anabolic and catabolic reactions in plants and animals of the sulphur amino acids and the other compounds listed in Table 3. This subject of sulphur biochemistry has been adequately covered in many reviews and texts (see above). However, for completeness a chapter has been included in this volume on naturally occurring sulphur compounds (Chapter 2 in Volume 1, Part A), and a brief overview is provided below.

Table 3 — List of some important sulphur compounds[a] derived from sulphur amino acids

Cysteine $\rightleftharpoons$ Cystine	Methionine
Proteins	Proteins
Glutathione	Cysteine
Coenzyme A	S-Adenosyl-L-methionine (SAM)
Cysteamine, taurine	Thiamine
Biotin	Lipoic acid (?)
Cysteinylflavins	Decarboxylated SAM[b]
Sulphate[c]	Sulphate[c]

[a] Only *primary sulphur metabolites* are listed in this table.
[b] Decarboxylated SAM is used in the biosynthesis of spermidine.
[c] Sulphate is activated to PAPS and used in the biosynthesis of sulphomucopolysaccharides and other sulphuric acid esters.

1.4.2   Sulphur biochemistry — an overview

The pools of oxidized sulphur (e.g. sulphate, SO_4^{2-}) and reduced sulphur (e.g. thiol, $-SH$, or thioether, $-S-$) in the biosphere can be interconverted in organisms. The special chemistry of sulphur (see section 1.3) allows these oxido-reductive changes to occur with relative ease, affording a diversity of products in nature. In animals and certain microorganisms divalent sulphur (oxidation state -2) metabolism is attended by a movement towards a higher oxidative state (SO_4^{2-}, $+6$). In plants and some microorganisms, the reverse occurs, i.e. sulphate is taken up and converted to organic (divalent) forms, primarily cysteine, cystine and methionine. Whereas part of the animal organic sulphur is returned to the biosphere as inorganic sulphate in excreta, the bulk of this sulphur is only recovered after death. Bacterial action in dead plants and animals liberates hydrogen sulphide, which then undergoes enzymic and non-enzymic conversion to sulphate. These reactions taken together constitute the 'sulphur cycle' (Fig. 1). The biotransformations of terrestrial elemental sulphur can also be incorporated in this cycle. The conversion of sulphur to sulphide occurs widely by non-enzymic reactions, but certain microbes are also able to effect this conversion. There is currently a renewed interest in microbiological sulphur cycling, and it has been suggested that sulphate reduction and photolithotrophic sulphur oxidation represent primitive forms of respiration and photosynthesis which developed on an anaerobic planet prior to the evolution of the H_2-splitting reaction of green plant photosynthesis (see Siegel, 1975, and references cited therein). Therefore energy metabolism based on sulphur may well have preceded that based on oxygen by about 1500 million years. Ancestor tracing based on analyses of DNA sequences concludes that all living organisms originated in a sulphur-metabolizing thermophilic organism (Penny, 1988; Lake, 1988).

The ultimate product of organic sulphur metabolism is sulphate (see above and Fig. 1). However, within organisms a wide variety of compounds are generated as intermediary metabolites (Table 3), many of which play critical roles in cellular biochemistry. A discussion of the biosynthesis, utilization and degradation of such

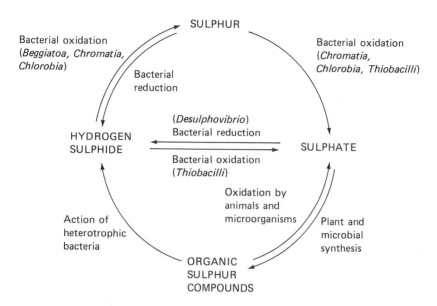

Fig. 1 — The sulphur cycle in nature (from Young and Maw, 1958). Reproduced with permission of Metheun & Co. Ltd.

organosulphur is beyond the scope of this chapter, and the reader is referred to biochemistry texts for this information. However, it is worth reiterating that the reactions these compounds participate in are absolutely reliant on certain aspects of sulphur functional group reactivities. For example the reactivity of the thiol (−SH) group in cysteine is pertinent to its role in structural and enzymic proteins. The thiol groups of two non-adjacent cysteine units in proteins may be oxidized to produce a disulphide bond; such disulphide bridges of proteins contribute significantly to the tertiary structure of proteins. Another aspect of the chemistry of cysteine is that the thiol group is reactive towards various reagents, e.g. Hg^{2+}; *N*-ethylmaleimide, *p*-chloromercuribenzoate. Since a free −SH group is vital for the catalytic function of many enzymes, such reagents destroy or inhibit many such enzymes. The reactivity of the cysteine −SH group in glutathione is important in its ability to scavenge and detoxify reactive electrophilic intermediates produced in cells. Similarly, the biological importance of coenzyme A lies in its ability to react via the −SH group with carboxylic acids to the corresponding thioesters. Methionine is the major methyl group donor in most biological transmethylating systems. Methionine is activated to *S*-adenosyl-L-methionine, a sulphonium compound readily susceptible to C–S$^+$ bond fission via nucleophilic displacement reactions. It has been suggested that other amino acid derivatives or other 'onium' compounds would not adequately serve this role as methyl group donors; quaternary ammonium compounds are too thermodynamically stable to methylate most acceptors effectively, and oxonium compounds (e.g. a hypothetical oxygen analogue of *S*-adenosyl-L-methionine) lack the kinetic stability to survive *in vivo* (Griffith, 1987).

It is hoped that the above overview of sulphur biochemistry puts the subject of sulphur xenobiochemistry into a proper context. As will be evident from section 1.5, and from subsequent chapters in these volumes on sulphur, chemical aspects are also important with sulphur xenobiotics, and many of their metabolic pathways, are governed by the chemical reactivities of the sulphur functional group they possess.

1.5 ASPECTS OF SULPHUR XENOBIOCHEMISTRY

1.5.1 Definition of sulphur xenobiochemistry
Diverse sources contribute to the environmental burden of non-physiological organosulphur compounds. Sulphur-containing compounds are used extensively as pesticides and as intermediates in the chemical industry. Others are also present in vast quantities in crude oil and other fossil fuels. In mammals that consume plant and microbial products, the 'sulphur secondary metabolites' (see section 1.4.1) must also be regarded as exogenous sulphur compounds. The use of organosulphur compounds as medicinal agents (see Chapter 4 of Volume 1, Part A) constitutes the intentional exposure of man to this class of chemicals. This requires that we have a clear understanding of the effect and fate of sulphur compounds in various living systems.

Sulphur xenobiochemistry may be defined as the study of the metabolism of exogenous sulphur compounds in living systems, i.e. a study of their absorption, distribution, biotransformation and elimination from organisms, and the study of the molecular basis of their pharmacological and/or toxicological effects. In such a broad definition, one may therefore include the study of the microbial and plant metabolism of sulphur pesticides at one end of the spectrum, and mammalian metabolism of organosulphur drugs and xenobiotics at the other. Many of these aspects are dealt with in detail in subsequent chapters in this volume.

1.5.2 Metabolic options open to sulphur xenobiotics
As in the case of most other xenobiotics, organosulphur compounds undergo a variety of phase I and phase II metabolic reactions, depending on the type of sulphur–containing functional groups in the molecule (Table 4). The most common metabolic pathway for sulphur xenobiotics appears to be *S*-oxidation. This is not surprising in view of the known chemistry of divalent sulphur functional groups (thiol, sulphide) (see section 1.3). The lone pair of electrons on divalent sulphur are readily accessible and can therefore participate in many reactions. *S*-Oxidation affords either *S*-oxygenated metabolites (sulphoxides, sulphones, sulphenic acids, sulphinic acids, sulphonic acids) or disulphides (probably via reactive *S*-oxygenated derivatives of thiols). Reductions have also been reported, e.g. with sulphoxides and disulphides. If one views sulphonium ion formation as a two-electron oxidation, then conversion of the sulphonium ion back to the sulphide must be regarded as a reduction. Reduction of sulphone functionalities is rare and still somewhat controversial (see Chapter 5 of Volume 1, Part B) Hydrolytic reactions occur with sulphamates (rarely) and most commonly with sulphate esters.

Carbon–sulphur bond formation occurs with sulphides, thiols, dithioic acids and thiocarbamides, because all of these can undergo conjugation reactions directly at their sulphur centres. *S*-Methylation of thiol groups, particularly aryl thiols, is common and is followed by *S*-oxygenation to water-soluble sulphoxides and sul-

Table 4 — Metabolic options open to sulphur functionalities

Sulphur functionality	Potential metabolite(s)	Reaction type
Sulphide R–S–R¹	Sulphoxide $\overset{O}{\underset{\parallel}{R\!-\!S\!-\!R^1}}$	Oxidation (sulphoxidation)
	Sulphone $O\!=\!S\!-\!R^1,\ =\!O$	Oxidation (sulphoxidation)
	Sulphonium ion $\underset{CH_3}{R\!-\!S^+\!-\!R^1}$	Conjugation (*S*-methylation)
	Thiol R–SH	Oxidation (*S*-dealkylation)
Sulphoxide $\overset{O}{\underset{\parallel}{R\!-\!S\!-\!R^1}}$	Sulphide R–S–R¹	Reduction (sulphoxide reduction)
	Sulphone $O\!=\!S\!-\!R^1,\ =\!O$	Oxidation (sulphoxidation)
Sulphone $\overset{O}{\underset{\parallel}{R\!-\!S\!-\!R^1}}\!=\!O$	Sulphoxide (?) $O\!=\!S\!-\!R^1$	Reduction (sulphone reduction)
	Sulphide (?) R–S–R¹	Reduction (sulphone reduction)

Substrate	Product name	Structure	Reaction
Thiol $R-SH$	Sulphenic acid	$R-SOH$	Oxidation (S-oxygenation)
	Sulphinic acid	$R-SO_2H$	Oxidation (S-oxygenation)
	Sulphonic acid	$R-SO_3H$	Oxidation (S-oxygenation)
	Disulphide	$R-S-S-R$	Oxidation (disulphide linkage)
	Mixed-disulphide (R^1SH endogenous thiol)	$R-S-S-R^1$	Oxidation (disulphide linkage)
	S-Thioacetyl	$R-S-COCH_3$	Conjugation (S-acetylation)
	S-Methylthiol	$R-S-CH_3$	Conjugation (S-methylation)
	S-Glucuronide	$R-S-C_6H_9O_6$	Conjugation (S-glucuronidation)
	S-Glucoside	$R-S-C_6H_{11}O_5$	Conjugation (S-glucosidation)
	S-Sulphate	$R-SO_3^-$	Conjugation (S-sulphation)
Disulphide $R-S-S-R^1$	Thiol	$R-SH + R^1-SH$	Reduction (disulphide reduction)
	Sulphoxide (?)	$R-\overset{O}{\underset{\parallel}{S}}-S-R^1$	Oxidation (sulphoxidation)
Dithioic acid $R-\overset{S}{\underset{\parallel}{C}}-SH$	'Sulphenic acid'	$R-\overset{S}{\underset{\parallel}{C}}-SOH$	Oxidation (S-oxygenation)
	'Sulphinic acid'	$R-\overset{S}{\underset{\parallel}{C}}-SO_2H$	Oxidation (S-oxygenation)
	S-Glucuronide	$R-\overset{S}{\underset{\parallel}{C}}-S-C_6H_9O_6$	Conjugation (S-glucuronidation)
	S-Glucoside	$R-\overset{S}{\underset{\parallel}{C}}-S-C_6H_{11}O_5$	Conjugation (S-glucosidation)
	S-Methyl	$R-\overset{S}{\underset{\parallel}{C}}-S-CH_3$	Conjugation (S-methylation)

Thioamide

$$R-\overset{\overset{\displaystyle S}{\|}}{C}-NH_2 \quad \left(\underset{R-C=NH}{\overset{SH}{|}} \right)$$

S-Oxide (sulphine)

$$R-\overset{\overset{\displaystyle S\rightarrow O}{\|}}{C}-NH_2$$

Oxidation (sulphoxidation)

di-S-Oxide (sulphene) (sulphene)

$$R-\overset{\overset{\displaystyle O\leftarrow S\rightarrow O}{\|}}{C}-NH_2 \quad \left(\underset{R-C=NH}{\overset{SO_2H}{|}} \right)$$

Oxidation (sulphoxidation)

Amide

$$R-\overset{\overset{\displaystyle O}{\|}}{C}-NH_2$$

Oxidation (desulphuration)

Phosphorothionate

$$R-\overset{\overset{\displaystyle S}{\|}}{\underset{R^1}{P}}-OR^2$$

Phosphate

$$R-\overset{\overset{\displaystyle O}{\|}}{\underset{R^1}{P}}-OR^2$$

Oxidation (desulphuration)

Thiocarbamide

$$R-N-\overset{\overset{\displaystyle S}{\|}}{C}-NHR^1 \quad \left(\underset{R-N-C=NR^1}{\overset{SH}{|}} \right)$$

Sulphenic acid

$$\underset{R-N-C=NR^1}{\overset{SOH}{|}}$$

Oxidation (S-oxygenation)

Sulphinic acid

$$\underset{R-N-C=NR^1}{\overset{SO_2H}{|}}$$

Oxidation (S-oxygenation)

Amide

$$R-N-\overset{\overset{\displaystyle O}{\|}}{C}-NHR^1$$

Oxidation (desulphuration)

S-Methyl

$$\underset{R-N-C=NR^1}{\overset{S-CH_3}{|}}$$

Conjugation (S-methylation)

Substrate	Metabolite	Structure	Reaction
Thiocarbamates (carbamothioates) $O=\;R-S-C-NR^1R^2$	S-Glucuronide	$R-N-C=NR^1$ with $S-C_6H_9O_6$	Conjugation (S-glucuronidation)
	Sulphoxide	$R-S-C-NR^1R^2$ ($O\;O$)	Oxidation (sulphoxidation)
	Glutathione cleavage products	$R-SOH+GS-C-NR^1R^2$ ($O=$)	Conjugation (glutathione conjugation)
Sulphonamide $R-SO_2NH_2$ (R = 2-imidazole, or 2-thiazole)	Glutathione conjugate	$R-SG$	Conjugation (GSH-dependent displacement of SO_2NH_2)
Sulphonium $R-S^+-R^2$ $\;|\;R^1$	Sulphide	$R-S-R^1$	Reduction (C–S^+ bond cleavage)
Sulphamate $R-NH-SO_3H$	Amine	$R-NH_2$	Hydrolysis
Sulphonate $R-SO_3H$	Sulphonic acid group is largely metabolically inert Glutathione conjugate $R-SG$		Nucleophilic displacement of alkyl sulphonates ?
Sulphate $R-OSO_3H$	Alcohol or phenol	$R-OH$	Hydrolysis

phones. *S*-Methylation has also been reported in dithioic acids and thiocarbamides. An interesting recent example of *S*-methylation is that of thioethers (sulphides); this novel pathway affords methyl sulphonium ions as metabolites (see Chapter 6 of Volume 1, Part A). *S*-Glycosylation of thiols, dithioic acids and thiocarbamides are other examples where C–S bonds are formed through conjugation of sulphur with glucuronic acid or glucose.

The role of *S*-acetylation reactions in the metabolism of mercaptans (thiols) is uncertain. Acetylation of various aliphatic thiols, by an acetyl-CoA-dependent enzyme, has been described *in vitro* (see Chapter 6 of Volume 1, Part A), but there are no examples of such reactions *in vivo*. There are very recent reports of novel sulphotransferase-mediated sulphate conjugation of xenobiotic thiols. Such compounds are apparently highly reactive and unstable, and their role in the *in vivo* disposition of thiols is not clear. Thiocarbamates and thiazole- and imidazole-2-sulphonamides undergo interesting reactions with glutathione, which are consistent with the known chemical reactivities of those functionalities. Thiocarbamates are initially oxidized to the corresponding sulphoxides, which are extremely reactive in the presence of excess glutathione, affording two cleavage products (see Chapter 4 of Volume 1, Part B). A similar C–S bond cleavage also occurs with the heterocyclic sulphonamides; the 2-sulphonamido group is readily displaced by glutathione, which is consistent with their reported chemical reactivities (Stirling, 1974).

Carbon–sulphur bond fission occurs during oxidative desulphurations and *S*-dealkylations. The former involves initial oxygenation at sulphur to produce oxo-derivatives which are thought to be intermediates in the desulphuration process. *S*-Dealkylations on the other hand involve an initial oxidative attack at the α-carbon; the unstable hydroxyalkyl derivatives are then converted to thiols and the corresponding aldehydes.

1.6   CONCLUDING REMARKS

From the brief comments in the section above, and from data presented in Table 4, it is apparent that the metabolic changes undergone by sulphur xenobiotics are many and varied in character. This will be emphasized in each of the subsequent chapters in this volume, on metabolism of sulphur functional groups. Depending on the reaction conditions, and on the structure of the sulphur functionality, sulphur xenobiotics can undergo oxidations, reductions, hydrolyses, or carbon–sulphur bond formation (conjugations) or carbon–sulphur bond fission (*S*-dealkylations, desulphurations). As in the case of endogenous compound reactivities (see section 1.4.2), biotransformations of foreign compounds are reliant on the special chemistry of sulphur. If one takes into account these chemical reactivites and physicochemical properties, one can often make reasonable predictions of the likely metabolic profiles, or at least rationalize the observed biotransformations.

REFERENCES

Anderson, J. W. (1978). In *Sulphur in Biology*, Edward Arnold Ltd., London, pp. 1–6.

Baudrimont, A. and Malaguti, A. (1837). 'Recherches sur la cystine' *C. R. Acad. Sci.*, **5**, 394–399.

Bixby, D. W. (1978). In *The Role of Phosphorus in Agriculture*, American Society of Agronomy, Madison, USA, pp. 125–132.

Burns, J. J., Yu, T. F., Ritterband, A., Perel, J. M., Gutman, A. B. and Brodie, B. B. (1957). Potent uricosuric agent: sulphoxide metabolite of the phenylbutazone analog G-25671. *J. Pharm. Exp. Ther.*, **119**, 418–426.

Clare, N. T. (1947). A Photosensitized keratitis in young cattle following the use of phenothiazine as an anthelmintic. II. The metabolism of phenothiazine in ruminants. *Aust. Vet. J.*, **23**, 340–344.

Cotton, F. A. and Wilkinson, G. (1972). *In*, 'Advanced Inorganic Chemistry: A Comprehensive Text', 3rd edition., Wiley, New York, USA, pp. 421–426.

Debus, H. (1850). Über die Verbindungen der sulphocarbaminsäure. *Ann. Chem. (Liebigs)*, **73**, 26–30.

deRey-Pailhade, J. (1888). Sur un corps d'origine organique hydrogénant le soufre à froid. *C. R. Acad. Sci.*, **106**, 1683–1684.

Fischer, E. (1878). Über die Hydrazinverbindungen. *Ann. Chem. (Liebigs)*, **190**, 114–123.

Friedmann, E. (1902). Über die Konstitution des eiweifscystins. *Beitr. Chem. Physiol. Path.*, *2*, 433–434.

Friedmann, E. (1903). Über die Konstitution des cystins. *Beitr. Chem. Physiol. Path.*, *3*, 1–46.

Gillette, J. R. and Kamin, J. J. (1960). The enzymatic formation of sulfoxides: the oxidation of chlorpromazine and 4,4′diaminodiphenyl sulphide by guinea pig liver microsomes. *J. Pharmacol. Exp. Ther.*, **130**, 262–267.

Giusti, G. P. (1965). *Oil Gas J.*, **63**, 99–100.

Greenberg, D. M. (ed.) (1975). *In*, 'Metabolism of Sulphur Compounds', Volume VII of *Metabolic Pathways*, Academic Press, New York.

Griffith, O. W. (1987). Mammalian sulphur amino acid metabolism: an overview. In W. B. Jakoby and O. W. Griffith (eds.), *Methods in Enzymology*, Vol. 143, *Sulphur and Sulphur Amino Acids*, Academic Press, London.

Hanzlik, R. P. (1984). Prediction of metabolic pathways — sulphur functional groups. In J. Caldwell and G. D. Paulson (eds.), *Foreign Compound Metabolism*, Taylor & Francis Ltd., London, pp. 65–78.

Hendrickson, J. B., Cram, D. J. and Hammond, G. S. (1970). In *Organic Chemistry*, McGraw-Hill, New York, pp. 789–814.

Hopkins, F. G. (1929). Glutathione — A reinvestigation. *J. Biol. Chem.*, **84**, 269–320.

Jakoby, W. B. and Griffith, O. W. (eds.) (1987). In *Methods in Enzymology*, Vol. 143, *Sulphur and Sulphur Amino Acids*, Academic Press, London.

Jones, D. N. (ed.) (1979). In *Comprehensive Organic Chemistry: The Synthesis and Reactions of Organic Compounds*, Vol. 3, 'Sulfur, Selenium, Silicon, Boron, Organometallic Compounds, Pergamon Press, Oxford.

Kane, P. O. (1962). Identification of a metabolite of the antituberculous drug ethionamide. *Nature*, **195**, 495–496.

Kharasch, N. and Meyers, C. Y. (eds.) (1966). In *The Chemistry of Organic Sulphur Compounds*, Vol. 2, Pergamon, Oxford.

Lake, J. A. (1988). Origin of the eucaryotic nucleus determined by rate-invariant analysis of *r*RNA sequences. *Nature*, **331**, 184–186.

Lavoisier, A. L. (1783). Nouvelles réflexions sur l'augmentation de poids quacquièrent, en brûlant, le soufre & le phosphore; & sur la cause à laquelle on doit l'attribuer. *Mém. Acad. Ray. Sci.*, 416–422.

Liebig, J. (1834). Ueber die Darstellung des mercaptans und den schwefe lcyanäther. *Justus Liebigs Ann. Chem.*, **11**, 14–18.

Lundy, W. T. (1949). Sulphur in pyrites. In *Industrial Minerals and Rocks*, 2nd edn., AIME, New York, Chapter 7.

March, R. B., Metcalf, R. L., Fukuto, T. R. and Maxon, M. G. (1955). Metabolism of Systox in the white mouse and American cockroach. *J. Econ. Entomol.*, **48**, 355–363.

Mitchell, S. C. and Waring, R. H. (1986). The early history of xenobiotic sulphoxidation. *Drug Metab. Rev.*, **16**, 255–284.

Monier-Williams, M. (1899) In *A Sanskrit-English Dictionary* (First edition 1899), Oxford University Press, Oxford.

Mörner, K. A. H. (1899). Cystin, ein spaltungsprodukt der hornsubstanz. *Hoppe-Seyler's Z. Physiol. Chem.*, **28**, 595–615.

Mueller, J. H. (1922). A new sulphur-containing amino acid isolated from casein. *Proc. Soc. Exp. Biol.*, NY, **19**, 161–163.

Oae, S. (ed.) (1977). In *Organic Chemistry of Sulphur*, Plenum, New York.

Penny, D. (1988). What was the first living cell? *Nature*, **331**, 111–112.

Priestly, J. (1790). *Experiments and Observations on different Kinds of Air (Birmingham)*, **2**, 295.

Reid, E. E. (1963). In *Organic Chemistry of Bivalent Sulphur*, Vol. V, Chemical Publishing Co., Inc, New York, pp. 422–445.

Rose, F. L. and Spinks, A. (1948). Metabolism of aryl sulphides: Part I. Conversion of *p*-methylthioaniline to *p*-methylsulphonyl aniline in the mouse. *Biochem. J.*, **43**, vii.

Salzman, N. P., Moran, N. C. and Brodie, B. B. (1955). Identification and Pharmacological Properties of a Major Metabolite of Chlorpromazine. *Nature*, **176**, 1122–1123.

Scheele, C. W. (1777). *Chemical Treatise on Air and Fire*, (Upsala and Leipzig).

Siegel, L. M. (1975). Biochemistry of the sulphur cycle. In D. M. Greenberg (ed.), *Metabolic Pathways*, Vol. VII, *Metabolism of Sulphur Compounds*, Academic Press, New York, pp. 217–286.

Smith, W. J. (1892). Uber das Verhalten von Carbaminthiosäureäthylester und Thiocarbaminsäureäthylester. *Pflüg. Arch. ges. Physiol.*, **53**, 481–490.

Smith, W. J. (1893). Ueber das Verhalten einiger schwefelhaltiger Verbindungen im stoffwechsel. *Z. Physiol. Chem.*, **17**, 459–467.

Smith, W. J. (1894). Weiteres über die Schwefelsäure-Bildung im Organismes. *Pflüg. Arch. ges. Physiol.*, **57**, 418–426.

Snow, G. A. (1957). The metabolism of compounds related to ethanethiol. *Biochem. J.*, **65**, 77–82.

Stirling, C. J. M. (1974). The sulfinic acids and their derivatives. *Int. J. Sulphur Chem.*, **6**, 277–320.

Thaulow, M. J. C. (1838). Über die zusammensetzung des blasenoxyds (cystic-oxyd). *Justus Liebigs Ann. Chemie*, **27**, 197–201.

Underhill, F. P. and Classon, O. E. (1905). The physiological behavior of methylene blue and methylene azure: A contribution to be study of the oxidation and reduction processes in the animal body. *Am. J. Physiol.*, **13**, 358–371.

Walkenstein, S. S. and Seifter, J. (1959). Fate, distribution and excretion of Promazine-S^{35}. *J. Pharmacol. Exp. Ther.*, **125**, 283–286.

Wollaston, W. H. (1810). On cystic oxide, a new species of urinary calculus. *Phil. Trans. Roy. Soc.*, 223–230.

Young, L. and Maw, G. A. (1958) In *The Metabolism of Sulphur Compounds*, Methuen, London.

Zehnder, K., Kalberer, F., Kreis, W. and Ruthschmann, J. (1962). The metabolism of thiethylperozine. *Biochem. Pharmacol.*, **11**, 551–556.

Zeise, W. C. (1834). Uber das mercaptan. *Justus Liebigs Ann. Chem.*, **11**, 1–10.

2

Naturally occurring sulphur compounds

A. B. Hanley and **G. R. Fenwick**
AFRC Institute of Food Research,. Norwich Laboratory, Colney Lane, Norwich
NR4 7UA, UK

SUMMARY

1. Naturally occurring sulphur compounds comprise a range of disparate compounds.
2. Despite this structural catholicity, certain features of the chemical and physiological properties of sulphur compounds are consistent throughout.
3. Such similarities are particularly noticeable when considering the biosynthesis of naturally occurring sulphur compounds.
4. Similarities are observed between compounds as structurally diverse as the vitamins biotin and thiamine, lipoic acid and the antibiotics penicillin and cepahalosporin.

2.1 INTRODUCTION

While, perhaps, less well studied and understood than nitrogen-containing natural products, the role of sulphur compounds is nonetheless crucial in the biosphere. It is possible to classify naturally occurring sulphur compounds into two categories which are common to all natural products. The first group consists of those compounds considered as essential to the survival and growth of the organism in question, i.e. primary metabolites such as the sulphur amino acids, cysteine and methionine (and closely related biological species), sulphur-containing peptides and proteins such as glutathione and metallothionein and essential cofactors including biotin, thiamine and lipoic acid. The second group of sulphur metabolites are those considered to be secondary products — compounds which are either non-essential to life or whose biological role is unclear. Although a number of definitions have been proposed to encompass secondary products, the diversity of their role in nature makes subclassification difficult. Moreover, a number of products have been isolated which, although initially appearing to have no particular biological importance, were later found to be essential. The present authors would regard primary metabolites as those which are essential to life and common throughout nature. Secondary metabolites, thereby, are considered as those of restricted or eccentric occurrence. Since the importance of secondary metabolites is, in many instances, unclear a number of theories concerning their role have developed. They have been suggested as the following:

   (i)   a mechanism for removal of excess primary metabolites (although their separate biological activity *per se* may militate against this theory);
  (ii)   as a means of removing surplus intermediates;
 (iii)   a store of potential primary metabolites in a biologically inert or altered form;
 (iv)   perhaps most interestingly, important mediators between species (Haslam, 1985).

Particular members of this class include penicillins and cephalosporins, glucosinolates (a group of flavour precursors isolated from plants of the Cruciferae), alk(en)yl-L-cysteine sulphoxides fulfilling a similar role within the genus *Allium* and miscellaneous other sulphur-containing compounds, notably sulphur alkaloids and thioketones.

Since the flavour precursors of the Cruciferae family and genus *Allium* together with asparagusic acid from asparagus are dealt with elsewhere in this volume (Part B, chapter 10), this chapter will be concerned primarily with the essential sulphur-containing compounds mentioned above and, in addition, sulphur-containing antibiotics. Particular emphasis will be placed on the biological, biochemical and chemical interrelationships and on similarities between the groups of compounds considered. The physiological properties of the compounds, while referred to, will not be treated in detail, since these are the subject of considerable and comprehensive coverage. Equally, a fully comprehensive survey is impossible in such a short space and interested readers are advised to consult the key references for each group of compounds.

The biological role of organosulphur compounds is a direct consequence of the chemical characteristics of the carbon–sulphur bond. The lack of stability of the thiolester in acetylcoenzyme A compared with the corresponding ester makes the former an efficient acyl group donor. The same argument applies for methylation reactions carried out by methionine. The biological activity of biotin is certainly dependent on the ring sulphur atom — possibly via a doubly protonated intermediate (*vide infra*). The ability of organothiol groups to revert reversibly to the oxidized form which gives rise to disulphide bridges is essential in the biological reduction of sulphate, in the biosynthesis of cysteine and methionine as well as in the structure and conformation of proteins. Reversible acylation is an important property of cysteine in enzymes. Sulphur, therefore, serves as an activator in biological reactions, making many reactions energetically favourable and thus biologically feasible.

2.2 SULPHUR COMPOUNDS AND THE SULPHUR CYCLE

The cycling of sulphur occurs in a similar fashion to that of nitrogen and a range of biological hosts are involved. Plants and many microorganisms derive necessary organic sulphur compounds from sulphate (Fig. 1).

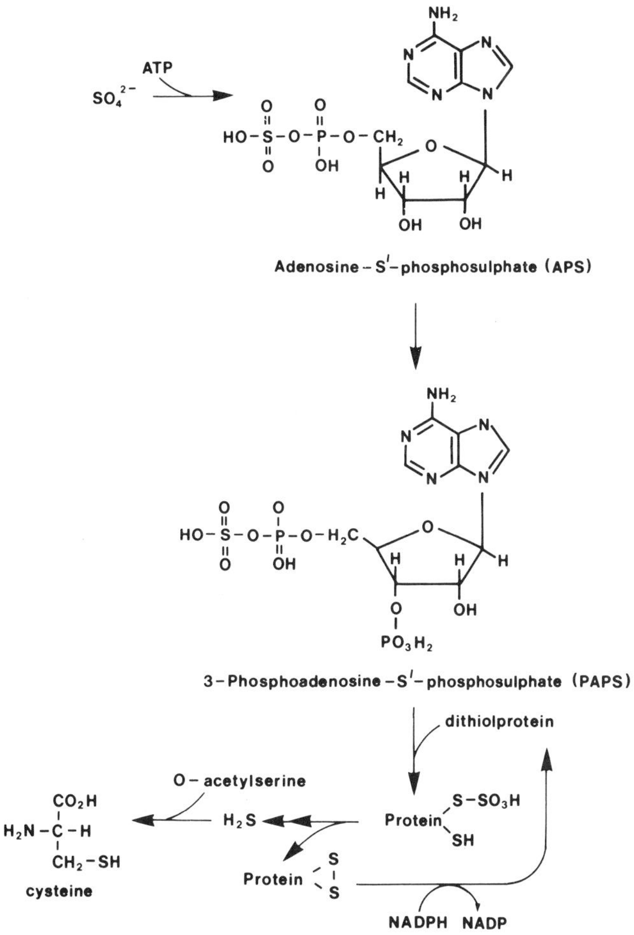

Fig. 1 — Plant and microbial reduction of sulphate to organosulphur compounds.

Sulphate is added to ATP to give adenosine-5′-phosphosulphate (APS) which is then phosphorylated to 3′-phosphoadenosine-5′-phosphosulphate (PAPS). Reaction with a dithiol protein gives a conjugate which then decomposes giving bisulphite and the oxidized protein. Sulphite reductase (Lazzarini and Atkinson, 1961; Tamura *et al.*, 1967; Naiki, 1965) then reduces the bisulphite to H_2S which, in turn, can react with *O*-acetylserine to give cysteine. Conjugates of homoserine (*O*-succinyl, *O*-acetyl and *O*-phospho) can also react with H_2S to give homocysteine in certain species (Giovanelli and Mudd, 1967; Datko *et al.*, 1977). The intermediate PAPS can also sulphonate suitable hydroxyl groups *in vivo*.

Other minor pathways also occur including the reversible addition of H_2S to serine in yeast (Schlossman and Lynen, 1957) and the use of sulphate metabolites as terminal electron acceptors in the oxidation of organic molecules in certain obligate anaerobes (Cooper, 1983).

2.3 SULPHUR-CONTAINING AMINO ACIDS

The biochemistry of sulphur-containing amino acids has been reviewed (Cooper, 1983).

2.3.1 Cysteine

The major source of bio-organic sulphur is via the sulphur-containing amino acids — particularly cysteine, methionine and closely related compounds. In animals, cysteine is derived from methionine and serine via the intermediate cystathionine (Cooper, 1983). In bacteria the reverse is the case with methionine being produced via cysteine and homocysteine (see Fig. 2) (Giovanelli *et al.*, 1978; Manicoll *et al.*,

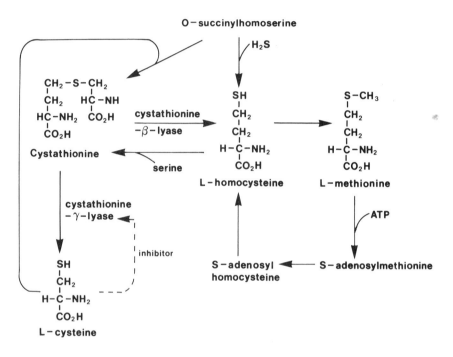

Fig. 2 — Cycling of cysteine and methionine.

1981). In both pathways the key intermediate is cystathionine which is cleaved either with a cystathione-γ-lyase (animals) (Flavin and Slaughter, 1971) to give cysteine, α-oxobutyrate and ammonia or with a cystathionase-β-lyase (E.C. 4.4.1.9) (microorganisms) (Giovanelli and Mudd, 1971) to provide homocysteine, pyruvate and ammonia and, after methylation, methionine. Both cystathionine and cysteine and their seleno analogues are substrates for an enzyme which mediates an α-β elimination giving homocysteine, pyruvate and ammonia and is found in plant extracts (McCluskey *et al.*, 1986). In certain microorganisms, H_2S is reacted directly with serine, mediated by cysteine synthase to give cysteine (Kredich *et al.*, 1969). The transulphuration pathway thus acts as a shunt for cycling homocysteine and methionine and maintaining the correct level of methionine *in vivo*. Cysteine synthase itself is a highly adaptable enzyme able to synthesize a number of 5-substituted cysteines from O-acetyl-L-serine and thiols (Murakoshi *et al.*, 1986; Murakoshi and Kaneko, 1985).

While both cysteine and homocysteine are readily oxidized to cystine, the intracellular cysteine:cystine ratio is high (Crawhall and Segal, 1967) with cystine being rapidly converted to cysteine *in vivo* (Singer, 1975). Cysteine is a biosynthetic precursor of both glutathione (see below) and taurine, functions as an important constituent of certain proteins, is the source of sulphur for several thionucleotides which occur as minor components in RNA and DNA (Carbon *et al.*, 1965) and is the biosynthetic precursor of coenzyme A.

The degradation of cysteine has been extensively studied (Cooper, 1983) and can only be briefly dealt with in the space available. Two major pathways are known in mammals — direct oxidation to cysteine sulphinate which, in turn, is metabolized either to pyruvate or via hypotaurine to taurine or transamination of 3-mercaptopyruvate and then pyruvic acid. 3-Mercaptopyruvate has been shown to exist in solution in equilibrium with a cyclic dithiane (Cooper *et al.*, 1982). The reactivity of 3-mercaptopyruvate is greater than that of most α-keto acids and it readily undergoes aldol condensations (Cooper *et al.*, 1982).

2.3.2 Methionine

The principal biological importance of methionine is as a methyl donor via its activated derivative *S*-adenosyl-L-methionine. A specific methyltransferase mediates further reaction to give the required biologically methylated products and *S*-adenosylhomocysteine. The overall reaction in prokaryotes is shown in Fig. 3 (also see chapter 6, volume 1, Part B, and chapters 6, and 7 in volume 2, Part A of this series).

Methionine is a biosynthetic precursor of ethylene via 1-aminocyclopropane-l-carboxylate (Mapson, 1969). Both methionine and its α-oxo derivative are readily hydrolysed to give methanethiol and this pathway is responsible for the toxicity of methionine, particularly if the transsulphuration pathway functions inefficiently or not at all (Steele *et al.*, 1979; Dixon and Benevenga, 1980).

Whilst mammalian cells in culture can be grown in the presence of homocysteine rather than methionine (Kreis, 1979), some malignant mammalian cells have been shown to have an absolute requirement for methionine and growth can be halted by addition of L-methionase to the medium (Kreis *et al.*, 1980). The potential for use of enzymes in a pharmacological context remains an interesting possibility; however, the clearance time and susceptibility to hydrolysis suggest that *in vivo* inhibitors of enzyme activity may offer a better solution in this case.

Methionine is also of crucial importance in protein synthesis. The sequence of bases coding for methionine is the initiation codon for protein synthesis from t-RNA. Both methionyl-t-RNA and its active derivative *N*-formylmethionyl-t-RNA have been extensively investigated and interested readers are referred to the review by Kisselev and Favorova (1974). It is noteworthy that a methionine residue in protein can be uniquely cleaved by treatment with cyanogen bromide. This is a direct consequence of the presence of sulphur and the orientation of the peptide bond with respect to the disulphide linkage (see Fig. 4). While a similar reaction cannot occur

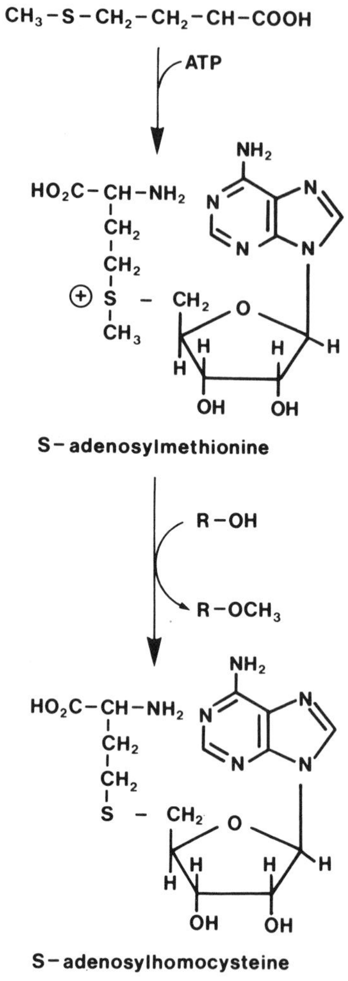
CH3-S-CH2-CH2-CH-COOH
ATP
NH2
HO2C-CH-NH2
CH2
CH2
S - CH2
CH3
N
O
H
OH
S-adenosylmethionine
R-OH
R-OCH3
S-adenosylhomocysteine

Fig. 3 — *In vivo* methylation with methionine.

with *S*-methylcysteine (Fig. 5), an oxazolinium bromide can be formed if the amino group is acylated. Subsequent rearrangement leads to ester hydrolysis. Although orientation is important, cystathione, which bears a partial structural similarity to methionine, does not react with cyanogen bromide — perhaps because of the proximity of the amino group, interfering with cyanogen bromide attack.

A classic example of the use of the cleavage at methionine by cyanogen bromide is demonstrated by the preparation of somatostatin using genetic engineering (Itakura *et al.*, 1977). The protein was produced by cloning the synthetically produced gene with an extra methionine terminal residue codon fused to a β-galactosidase gene. Somatostatin was released, after isolation of the fused protein from transformed cells, by treatment with cyanogen bromide.

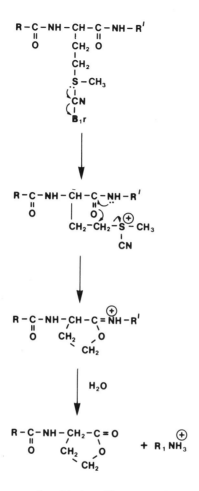

Fig. 4 — Cleavage of methionine with cyanogen bormidse.

The production of primary and secondary metabolites and proteins from cysteine and methionine will be considered in more detail elsewhere in this volume.

2.4 SULPHUR-CONTAINING VITAMINS AND COENZYMES

2.4.1 Coenzyme A

Coenzyme A, which was first isolated by Lippmann *et al.* (1947), is an acyl group carrier and is involved in fatty acid oxidation and synthesis, pyruvate oxidation and biological acetylations. The active carrier form is the thioester of acetic acid with the thiol group of coenzyme A (Fig. 6). The thioester is less stable than the corresponding ester and hence acts as a more efficient acy-group donor. While acetyl coenzyme A is formed as a result of the enzymic oxidation of pyruvate or fatty acids, it can also be produced from free acetate in a reaction mediated by acetylcoenzyme A synthetase.

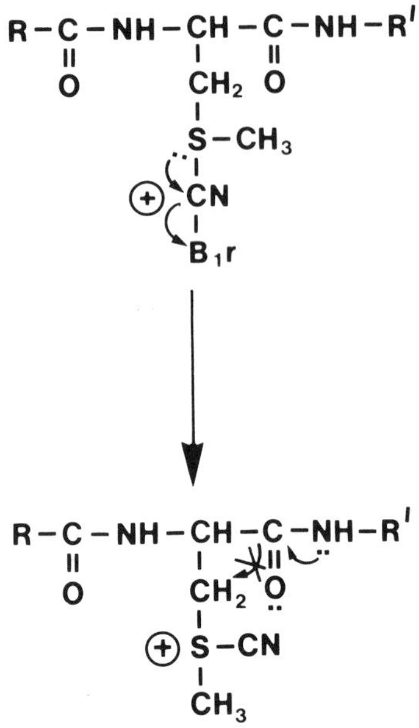

$$R-\underset{\underset{O}{\|}}{C}-NH-\underset{\underset{CH_2}{|}}{CH}-\underset{\underset{O}{\|}}{C}-NH-R^I$$

$$\underset{|}{S}-CH_3$$

Fig. 5 — Reaction of *S*-methylcysteine with cyanogen bromide.

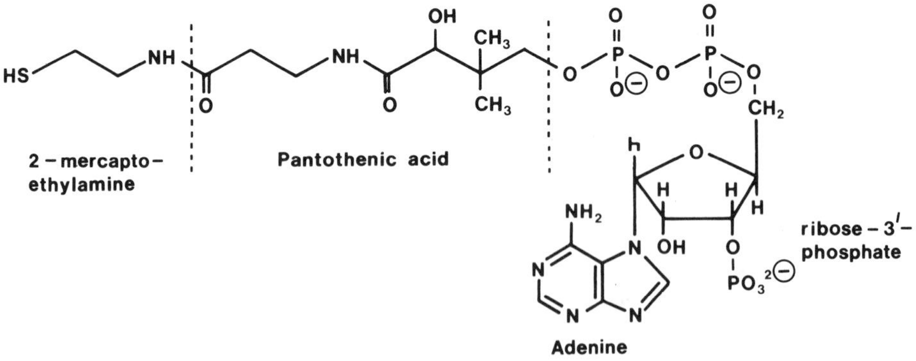

Fig. 6 — Structure of coenzyme A.

2.4.2 Lipoic acid

In common with many naturally occurring sulphur compounds, detailed investigation of lipoic acid was only carried out relatively recently. It was first isolated in a crystalline form by Reed *et al.* (1951); it had, however, been recognized since 1946 (Guirard *et al.*, 1946) as a growth factor and cofactor for pyruvate oxidation in microorganisms. Structural and synthetic investigations were carried out and the absolute configuration was determined by Mislow and Meluch (1956). Both α and β-lipoic acids show similar biological activity (Bullock *et al.*, 1954).

2.4.2.1 Biosynthesis

The biosynthesis of lipoic acid has been reviewed by Parry (1983). Reed (1966) first suggested that octanoic acid is a precursor of lipoic acid in *E. coli* This was later confirmed by Parry who studied the incorporation of C-5, C-6, C-7 and C-8 singly tritiated octanoic acid when fed simultaneously with ^{14}C-labelled octanoic acid (Parry, 1977). Measurements of ^{14}C:^{3}H ratios showed that C-6 and C-8 sulphur incorporation occurs without hydrogen loss of tritium from C-5 or C-7, thus eliminating the possibility of sulphur insertion resulting from unsaturation at C-5 or C-7 (Fig. 7). One hydrogen appears to be lost, stereospecifically, from C-6 and there

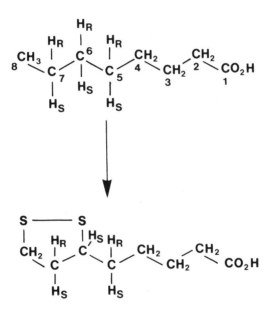

Fig. 7 — The biosynthesis of lipoic acid

is no loss from C-8. This latter result may be due to a substantial tritium isotope effect although further investigation is necessary. The feeding of stereospecifically tritiated 6-R and 6-S octanoic acid indicated that the 6 pro-R hydrogen atom is lost during biosynthesis (Parry and Trainor, 1978). Since the configuration of lipoic acid at C-6 is R (Mislow and Meluch, 1956), then sulphur insertion must occur with inversion or configuration. This is opposite to that considered to occur in biotin (see below).

The results of the radiochemical experiments were confirmed by studies of the incorporation of deuterium-labelled acetate (White, 1980a, c). The incorporation of up to eight deuterium atoms (as measured by mass spectrometry) suggested a fully saturated fatty acid as precursor. Administration of fully deuterated octanoic acid showed that 90% of the sample of isolated lipoic acid had been biosynthesized from the octanoic acid and that, furthermore, 13 deuterium atoms were incorporated — thus only two hydrogens are lost or replaced during the biosynthesis. The stereochemistry of biosynthesis was similarly confirmed by showing that the C-6 deuterium was retained from acetate-derived octanoic acid. Since it was known that a similarly derived hydrogen on C-10 of palmitic acid is incorporated in the pro-S position, then the pro-R rather than the pro-S hydrogen is lost at C-6, and thus sulphur is introduced with inversion of configuration.

In a further series of experiments using deuterated hydroxyoctanoic acids, White (1980b) demonstrated that thiooctanoic acids are incorporated into lipoic acid much more efficiently than the corresponding hydroxy derivatives, suggesting that the latter are not biosynthetic intermediates (Fig. 8). In particular, deuterated

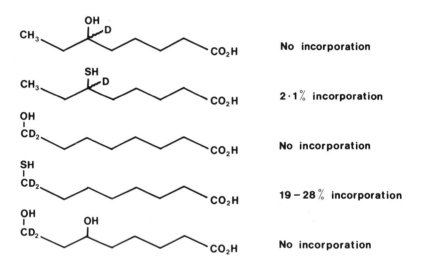

Fig. 8 — The biosynthesis of lipoic acid via thiol and oxygenated intermediates.

8-mercaptooctanoic acid is very effectively incorporated and, although Parry (1983) has been unable to detect the free thio acid in *E coli* by isotope dilution, it would appear to be an intermediate, although not as a free compound or else not accumulated in living tissue.

2.4.2.2   Future possibilities
The possibility of using genetic amplification or suppression of certain enzymes in the biosynthetic pathway to lipoic acid may offer a means for further studying this intriguing sulphur insertion reaction in detail. The occurrence of an apparent isotope effect at C-8 of octanoic acid warrants further detailed investigation. There are a

number of similarities between the biosynthesis of biotin and lipoic acid and these will be discussed in more detail below.

It is clear, particularly from the studies of White and coworkers, that stable isotope investigation of trace compounds such as lipoic acid is likely to prove increasingly important. The development of methods for the stereospecific synthesis of deuterated substrates and intermediates is clearly of prime importance — as is the continued development of mass spectrometric methods for the examination of those products of biosynthetic pathways which are produced in submilligram amounts.

2.4.3 Biotin

The discovery, structural elucidation and subsequent investigation of biotin (Fig. 9)

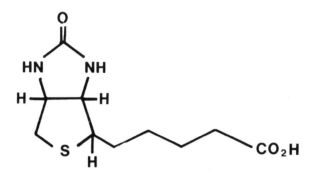

Fig. 9 — Structure of biotin.

mirrors closely advances in bio-organic chemistry. After the initial discovery in 1936 (Kogl and Tonnis, 1936), the structure was elucidated in 1942 (Melville *et al.*, 1942). However, it was not until 1958 that the role of biotin in carboxylation reactions began to be understood (Wakil *et al.*, 1958). A number of enzymes make use of biotin as a cofactor and these have been reviewed by Wood and Barden (1977). The overall reaction of biotin, a biotin enzyme and a substrate is considered to be as follows:

$$\text{E-biotin} + \text{ATP} + \text{HCO}_3 + \text{Mg}^{2+} \rightarrow \text{E-biotin-CO}_2^- + \text{ADP} + \text{P}_i \qquad (1)$$

$$\text{E-biotin-CO}_2^- + \text{substrate} \rightarrow \text{E-biotin} + \text{substrate-CO}_2^- \qquad (2)$$

A number of aspects of the catalysis by biotin enzymes are now known, and interested readers are referred to the review by Wood and Barden (1977). A brief summary of some of the most pertinent points will suffice for this chapter.

2.4.3.1 Biological action

Although the various biotin enzymes have wide quaternary structure variation, the mechanism of action is essentially similar. It is known that 1-*N'*-carboxybiotin is the biologically active form (Guchait *et al.*, 1974). Biotin is a very weak nucleophile and it has been the source of some speculation as to how it can be carboxylated effectively in the reverse of eq. (2), above (Mildvan and Fry, 1987). The possibility of the active species being an enol (De Titta *et al.*, 1976) has been suggested. It has also been proposed that a transannular interaction of the sulphur atom with the carbonyl carbon could be important (Mildvan *et al.* 1966). When the sulphur was replaced by oxygen, substantial loss in activity was noted (Lane *et al.*, 1964). However, IR and ^{13}C NMR investigation failed to detect significant transannular interaction (Bowen *et al.*, 1968). It has been found, however, that the N-1 proton exchanges with water protons at a rate comparable to the carboxylation rate of enzyme-bound biotin, implying that no additional effects are involved in the loss of the proton in the enolization of biotin (Fry *et al.*, 1985a, b). The side chain apparently hinders loss of the N-3 proton. A doubly-protonated biotin intermediate which is dependent on the presence of sulphur has been proposed (Fry *et al.*, 1985b). A mechanism involving transannular interaction of singly-protonated biotin has also been suggested and this would serve to increase the nucleophilicity at N-1. A more detailed description is provided by Mildvan and Fry (1987).

2.4.3.2 Biosynthesis
The biosynthesis of biotin has been extensively studied (Parry, 1983). Early work has been reviewed by Eisenberg (1973). These investigations have culminated in the biosynthetic pathway shown in Fig. 10. A number of aspects remain unclear, in

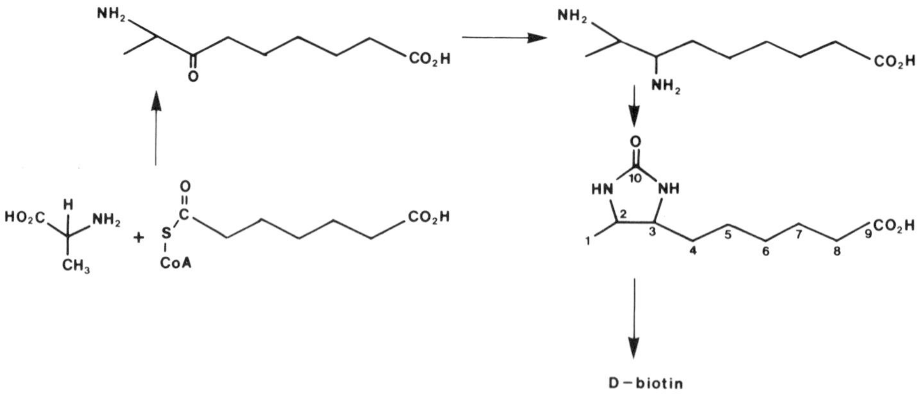

Fig. 10 — The biosynthesis of biotin.

particular the source and mechanism of sulphur insertion into desthiobiotin. Studies have, to a large extent, been hampered, as in the cases of thiamine and lipoic acid (*vide infra*), by the extremely small amounts of biotin produced *in vivo*. It should, however, be noted that Pearson *et al.* (1986) have assayed the production of biotin by yeasts and have found a strain of *Rhodotorula glutinis* capable of producing 10.6 ng per ml of culture filtrate.

Studies using stereospecifically labelled desthiobiotin in *Aspergillus niger* suggested that the sulphur insertion occurs without loss of hydrogen at C-2, C-3 or C-5

(Parry and Naidu, 1980); thus, analogously to the case of the lipoic acid, desaturation as an intermediate step is unlikely. Only (+)-desthiobiotin serves as a precursor. Furthermore, it was shown that one hydrogen is lost from C-1 non-stereospecifically while one is also lost from C-4 in a stereospecific fashion during the formation of biotin. In contrast with C-8 lipoic acid, no isotope effect was noted (Parry and Naidu, 1980). Similar studies in *E. coli* using deuterated and tritiated desthiobiotin indicated no hydrogen loss from C-2, C-3 or C-5 (Guillerm *et al.*, 1977). In addition [1-$^2H_3$]desthiobiotin was incorporated with loss of a single deuterim; thus the biosynthetic pathway in prokaryotes appears to be similar to that in eukaryotes (Frappier *et al.*, 1982). In a further elegant series of experiments, Parry and fellow workers have shown that sulphur insertion proceeds with loss of the 4 pro-S hydrogen (Trainor *et al.*, 1980) and that, unlike the case of lipoic acid, the reaction occurs with overall retention of configuration. A number of hydroxylated desthiobiotin intermediates have been tested (Frappier *et al.*, 1979); however, none of these (Salib *et al.*, 1979) would support growth of biotin-dependent strain of *E. coli*. A proposed intermediate, isolated in small amounts by Marquand (Salib *et al.*, 1979), has subsequently been shown not to be a true biosynthetic precursor (Ozaki, 1986). As with lipoic acid, the possibility of enzyme-bound intermediates cannot be excluded and further study using amplified strains would appear to be warranted.

2.4.4 Thiamine
2.4.4.1 Chemistry and biochemistry
Thiamine (vitamin B_1) is involved as a cofactor in biological decarboxylation reactions. It exists *in vivo* as the pyrophosphate which was first isolated from yeast by Lohmann and Schuster (1937). The mechanism of its action is shown in Fig. 11; all the reactions involve an acyl anion. The C-2 carbon can support a charge by virtue of its proximity to the positively charged nitrogen. In addition, C-2 is also a potent nucleophile. As well as being a cofactor in decarboxylations (although it should also be noted that thiamine can catalyse decarboxylations in the absence of any enzyme), it is also involved in a number of other reactions in which the nucleophilicity at C-2 and the stabilization of charge by the nitrogen are important. For example, hydroxyethylthiamine pyrophosphate (the *in vivo* form of the cofactor) can react with lipoic acid to give, initially, a hemithioacetal and, after decomposition, a thioester (Fig. 12).

The similarity between the role of thiamine in biochemical reactions and the catalytic effect of cyanide ion in benzoin condensations has been noted. Breslow (1958) has studied the anion formed at C-2 of the thiazole ring in thiamine by investigating the potential of thiamine and other thiazolium salts to act as catalysts in benzoin condensations. These studies have shown that the electron-withdrawing effect of the pyrimidine ring is an important factor as is the aromatic nature of the thiazolium ring. On the basis of these studies, he predicted that 2-acetylthiamine should be an effective acetylating agent and this proved to be the case. Thiamine has been attached to β-cyclodextrin and certain thiamine-type reactions are promoted without addition of exogenous enzyme. The substrate involved must be able to bind to the cyclodextrin cavity, and, while the thiamine-catalysed oxidation of aldehydes proceeds readily, the coupling of two aldehydes in a benzin-type condensation occurs less readily, owing to steric limitation.

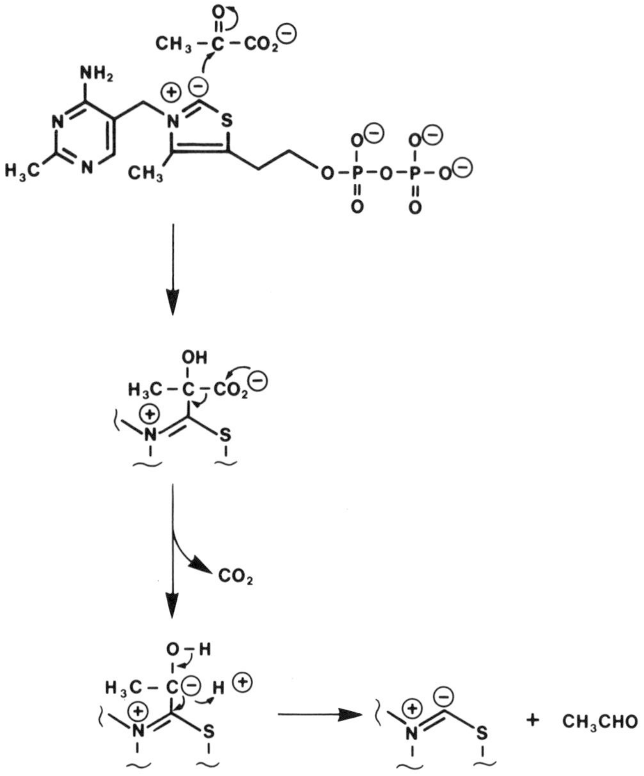

Fig. 11 — The biological activity of thiamine.

2.4.4.2 *Biosynthesis*

Although the biosynthesis of thiamine has been extensively studied, present knowledge remains incomplete — particularly with regard to the situation in eukaryotes. A number of reviews concerning the biosynthesis of thiamine have appeared recently and interested readers are referred to these (Young, 1986; Brown and Williamson, 1982; Leder, 1975).

Thiamine is composed of a pyrimidine moiety and thiazole ring. The coupling together of the two halves of the thiamine molecule was the first step in the biosynthetic pathway to be elucidated and the coupling enzyme has been isolated (Leder, 1970). Thiamine monophosphate, rather than thiamine itself, is the biosynthetic product. The coupling takes place between the pyrimidine pyrophosphate and thiazole phosphate and the former is synthesized two steps from the completely non-phosphorylated form. The thiamine phosphate formed must be dephosphorylated to give thiamine before being converted to the biologically-active form, thiamine pyrophosphate, by transfer of a pyrophosphate group from ATP (Shimazone *et al.*, 1959).

Thiamine is split into its two constituent molecules by two enzymes termed

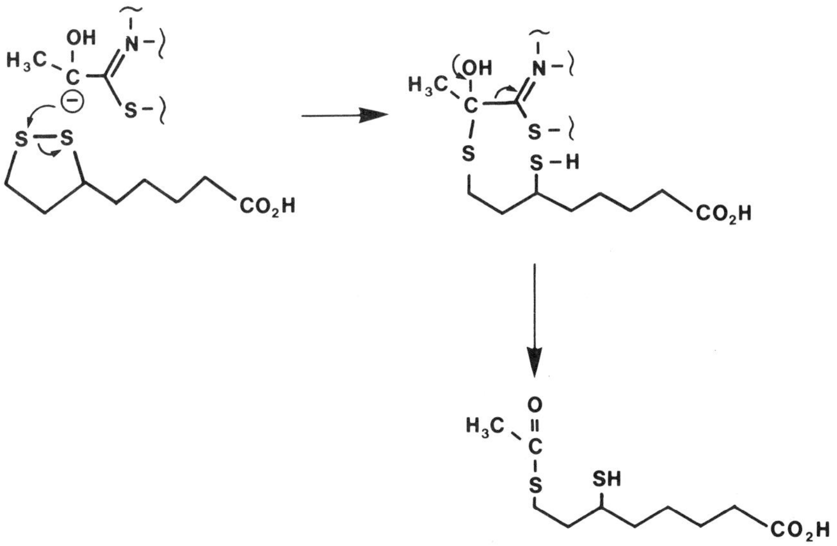

Fig. 12 — The reaction of thiamine with lipoic acid.

thiaminase I (E.C. 2.5.1.2) and thiaminase II (E.C. 3.5.99.2). The former requires a nucleophile (commonly an amine) and has been used as the basis for the investigation of the biosynthesis of thiamine using gas chromatography–mas spectrometry (Hanley, Baxter and Chan, unpublished) (Fig. 13) while the latter gives the 5-

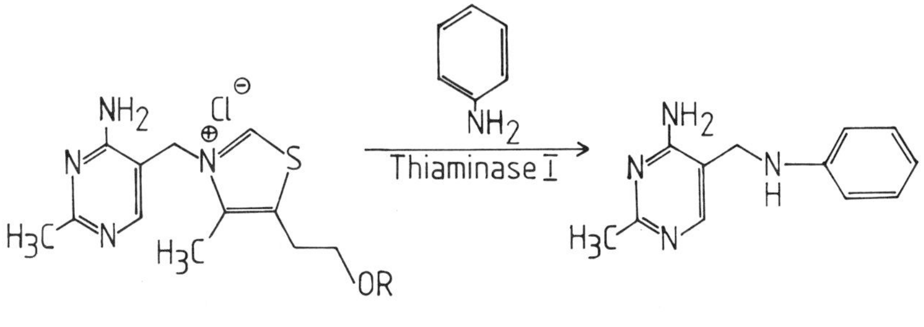

Fig. 13 — The use of thiaminase I to assay stable isotope incorporation in thiamin.

hydroxymethylpyrimidine derivative. The biosynthesis of the pyrimidine ring in bacteria was the subject of a series of elegant studies, particularly by Newell and Tucker (1967, 1968a, b), in which a common biosynthetic intermediate between thiamine and purines in prokaryotes, 5'-aminoimidazole ribonucleotide (5'-AIR), was confirmed. It is interesting to note that, although methionine is required for the conversion of 5'-AIR to the pyrimidine part of thiamine, there is no incorporation of methionine itself and its role in the process remains unclear (Newell and Tucker, 1986a, b). The situation in eukaryotes is less clear. Grue-Sørensen *et al.* (1986) have suggested that two pathways operate in *Saccharomyces cerevisae* with specific incorporation from labelled glucose and glycerol and the other carbon (either C-2 or C-4 depending on which pathway operates) being formate derived. Tazuya *et al.* (1986) in a similar series of experiments using *Candida albicans*, while finding the same precursors to be incorporated, found a different labelling pattern. The solution of this problem remains a major biosynthetic challenge. The same Japanese workers have recently proposed that the amide nitrogen of glutamine is incorporated in both *Saccharoimyces cerevisae* and in *E. coli* and Kozluk and Spencer, 1987 have found that, unlike the case in *E. coli*, purines and the pyrimidine molecule in thiamine do not have a common precursor in yeast.

The thiazole part of the molecule has also been extensively investigated. In *E. coli*, 4-hydroxybenzyl alcohol has been found to be a metabolite of tyrosine in *E. coli* grown in the absence of thiamine or thiazole (Therisod *et al.*, 1978) and, furthermore, [^{15}N]tyrosine is incorporated into thiazole (White and Rudolph, 1978). Evidence has accumulated which suggests a three-carbon sugar biogenesis of the carbon atoms in thiazole (White and Spencer, 1982; Yamada *et al.*, 1984). Differences occur between eukaryotes and prokaryotes with glycine and a pentulose phosphate being implicated in yeast while in *E. coli*, as mentioned above, tyrosine and a deoxypentulose are involved. The sulphur source is presumed to be a cysteine (Brown and Williamson, 1982) although other workers have suggested methionine (Hitchcock and Walker, 1961). The latter is unlikely and the results by these workers may have been due to the presence of a radioactive impurity in the isolated thiamine. Estramariex *et al.* have found a loss of incorporation of [^{35}S]sulphate into thiamine in a methionine auxotroph of *E. coli* when fed with glutathione or cysteine-neither methionine or homocysteine had an effect (Estramariex *et al.*, 1977). In a recent study DeMoll and Shive (1985) have shown that unlabelled methionine has no effect on labelled sulphate incorporation into thiamine; however, cysteine blocks incorporation indicating that cysteine is related to the sulphur in the thiazole ring in thiamine.

2.5 SULPHUR-CONTAINING PROTEINS

It is inappropriate to consider here the vast range of sulphur-containing proteins and their role *in vivo*. Instead we propose to concentrate on two biologically important molecules at differing ends of the size range, the tripeptide glutathione, and metallothionein comprising some 60 amino acid residues of which approximately one-third are sulphur containing. Finally the role of sulphur in enzymes and its importance to their activity, structure and function will be considered.

2.5.1 Glutathione
Glutathione (L-γ-glutamyl-L-cysteinylglycine) is generally the most common cellular thiol. After its initial discovery in 1888 (de Rey-Pailhade, 1888a, b) it was largely forgotten for over 30 years (Hopkins, 1921). It is involved in a plethora of biological reactions — functioning as a coenzyme, as a conjugate with xenobiotics and with

native toxins. The mechanism of action of glutathione-dependent enzymes has recently been reviewed (Douglas, 1987). Glutathione is synthesized intracellularly by the coupling of γ-glutamylcysteine to glycine — the reactions mediated by γ-glutamylcysteine synthase and glutathione synthase. The biological interrelationships of glutathione and other cellular metabolites are extremely varied and some of these are summarized in Fig. 14. The reaction types can be split in two — the

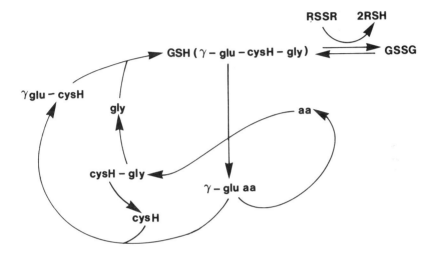

Fig. 14 — The biological interrelationships of glutathione.

oxidation–reduction cycle leading to interconversion of glutathione and glutathione disulphide and the γ-glutamyl cycle in which 5-oxoproline is involved (Meister and Anderson, 1983). The conversion of glutathione to glutathione disulphide is catalysed by a selenium-containing enzyme — glutathione peroxidase — or by transhydrogenases. The reverse reaction is carried out by glutathione disulphide reductase (Meister and Anderson, 1983).

Some indication of the biological role of glutathione is apparent from mutation studies carried out by Apontweil and Berends (1975a, b). In *E. coli* D12 mutants deficient in γ-glutamylcysteine synthetase or in glutathione synthetase, no glutathione could be detected. The mutants grew normally but both sulphydryl reagents and antibiotics had a pronounced effect. In a similar experiment, Fuchs and Warner (1975) studied the accumulation of γ-glutamyl cysteine in a glutathione synthase deficient mutant of *E. coli* and found levels similar to the expected amounts of glutathione in normal strains. Mutants lacking glutathione disulphide reductase grew normally with typical levels of glutathione, indicating the likelihood of an alternative reduction pathway.

Some of the differences in glutathione-related enzymes may prove of potential pharmacological importance. Many problems remain, however, in the biological action of glutathione and some of these are intimately linked with the major questions in bio-organic chemistry. In particular the transport, regulation, conjugation with amino acids and xenobiotics and its relevance to cell membranes would reward active investigation. The study of the differences between organisms that require glutathione and those which accumulate it but can grow without it (i.e. *E coli*) may provide some solutions.

2.5.2 Metallothionein

Metallothionein is a small, cysteine-rich protein involved in the binding of trace metals (Kagi and Nordberg, 1979). It occurs in a wide variety of eukaryotic species and, although small metal binding proteins have been found in prokaryotes, they appear to be different in form. Detailed investigation of metallothionein, including the use of molecular biological techniques, have shed some light on the role of this protein *in vivo* (Butt and Ecker, 1987). The protein exists in different types of varying complexity depending on the species. Cloning of the gene from yeast (*Saccharomyces cerevisiae*) was determined by the ability of the transformed cells to exhibit resistance to copper poisoning (Fogel and Welch, 1982). The copper metallthionine gene, CUP1, has been cloned into *Saccharomyces cerevisiae* using both high and low copy number vectors (Karin *et al.*, 1984). The synthesis of the protein can be induced by heavy metal poisoning as well as by glucocorticoids and interferon.

While significant studies have been carried out on genetic modification and variability of metallothionein, aspects of the bonding remains obscure. The metal-lothionein from yeast was first noted in 1975 (Prinz and Weser, 1975; Premakumar *et al.*, 1975). It is induced only by copper which is present as $Cu(I)$ and is ligated to 12 cysteines. Butt and Ecker (1987) have noted that the demetallated protein is susceptible to proteases and suggested that this is a manifestation of a lack of tertiary structure. Unlike mammalian metallothionein, the precise geometry of binding of the protein from yeast is obscure. It lacks significant structural homology with mammalian metallothionein and both X-ray crystallography and NMR examination are likely to be of use in structural determination. Studies on other metallothioneins (with the exception of that from mammalian systems) remain incomplete. The ability to clone into *Saccharomyces cerevisae* makes isolation of the protein in reasonable amounts a relatively simple procedure. A study of the binding of metals to the protein, selective deuteration of certain amino acids followed by NMR investigation and study of the protein in different environments may provide useful information regarding its function *in vivo*.

The precise biological role of metallothionein remains a matter for debate. While it is certainly involved in detoxification, the possibility of a specific metal transfer function has been speculated upon. Genetic modification of natural metallothionein may help to define, more strictly, the role and importance of the protein. The possibility of using modified metallothionein to mediate chemical reactions which have a strict orientation requirement and a need for $Cu(I)$ may merit further investigation. A more obvious use may be in the field of precious metal recovery.

2.5.3 Sulphur-containing enzymes

In addition to their role in conferring structural stability by virtue of disulphide linkages, sulphur-containing residues are important in enzyme activity *per se*. A few examples will serve to illustrate the types of reaction for which a sulphur ligand is important.

Zinc enzymes constitute a particularly important class of biocatalysts and, in many cases, sulphur amino acids (particularly cysteine) are of crucial importance. In liver alcohol dehydrogenase (E.C. 1.1.1.1), Zn^{2+} in the active site is coordinated to cysteine at positions 46 and 174 and to histidine 67 (Branden et al., 1975, Eklund et al., 1982). The binding of the cofactor NAD^+ causes a pronounced change in the active site with a five-coordinate Zn^{2+} being involved. The two cysteine residues lead to a higher than expected PK_a value for the water which is bound to Zn^{2+} in the active site. This leads to enhanced dehydrogenase activity.

Thiol proteases such as papain incorporate a cysteine residue into their active site. They occur in plants such as pineapple, papaya and kiwi fruit. Unlike serine proteases, hydrophobic amino acid sites are specifically cleaved, with cysteine-25 becoming acylated. The mechanism is considered to involve the attack of S- from cysteine on the substrate, generating a tetrahedral intermediate which then breaks down to give the cleaved product, with a histidine residue (156) probably acting as a general base (Fig. 15) (see Fersht, 1985, for a detailed discussion).

The continued interest in modification of enzyme activity makes the ability to manipulate enzymes and an understanding of the mechanism of action important. Clearly, replacement of, for example, serine with cysteine can modify activity without altering essentially the reaction which is being catalysed. The similarities between serine, threonine and cysteine makes them obvious candidates for interchanging in enzymes. Methionine, in contrast, has no obvious oxygen-containing analogue, although it bears some similarity to leucine and isoleucine It may be a useful means of fine tuning enzyme activity. It should also be noted that sulphur-containing residues in enzymes can be manipulated in a crude chemical sense by reaction with suitable reagents and ligands and this may also have a profound effect on enzyme activity and stability.

2.6 SULPHUR-CONTAINING ANTIBIOTICS OF MICROBIAL ORIGIN

The discovery of naturally-occurring sulphur-containing antibiotics can be said to have revolutionized the treatment of microbial infectious diseases. The efficacy of naturally-occurring sulphur compounds per se has been realized for some considerable time and indeed, as detailed elsewhere, was a major impetus behind investigation of the sulphur compounds of plants of the genus Allium (see chapter 10, this volume, Part B). It is proposed to discuss only those compounds of microbial origin. This means that the discussion will be limited principally to penicillins, cephalosporins and related species. In addition, because of the large number of excellent reviews and the continued high level of work in this area, this is not intended to be a detailed description of either the physiological effect of β-lactam antibiotics or, indeed, of many of the most recent advances in the chemistry of these compounds. Some mention will be made of the occurrence and general biological properties of sulphur-containing antibiotics and their biosynthesis (particularly with respect to the incorporation of sulphur), and some recent approaches to the modification of antibiotics will be briefly discussed.

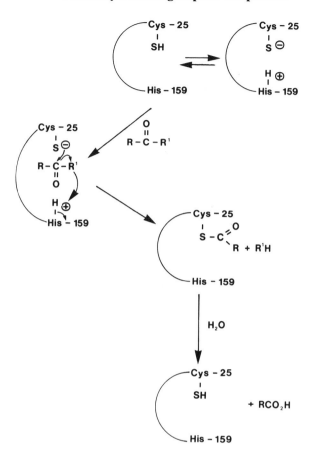

Fig. 15 — The activity of thiol proteases.

2.6.1 Occurrence and modification

While the isolation of pure penicillin and cephalosporin derivatives is a relatively recent occurrence, folk-lore suggests that crude culture mixtures have been used for some time. Unfortunately, precise description of the cultures and conditions of growth are generally not available, however, the following description by Cranch (cited in Selwyn, 1979) suggests that microbially-derived antibiotics were appreciated in the seventeenth century and earlier.

> For the last four generations there has been passed along in our family Thetrum Botanicum by John Parkinson, Apothecary of London and King's Herbalist. The very columinous work was published in 1640. In the fourteenth tribe of plants in which are included "marsh water and sea plants with mosses and mushrooms" we find the following among the descriptions of the various tree mosses:

Muscus ex Cranio humano.

The mosse upon dead mens sculles. Let me here also adjoyne, this kind of mosse somewhat like into the mosse of trees, and groweth upon the bare scalpes of men and women that have lyen long, and are kept in Charnelle houses in divers countries, which hath not only beene in former times much more set by, to make the Unguentum sympatheti-cum, which cureth wounds without locall application of salves, the composition where of is put as a principal ingredient, but as Crollius hath it, it should be taken from the sculles of those that have beene hanged or executed for offences.

It is interesting to consider whether there is any microbiological reason for these particular growth requirements. The tradition of applying mould to wounds has also been frequently commented on and, while the victim may eventually suffer from aflatoxin poisoning, the treatment was, in certain instances, efficacious.

Much of the early work on *Penicillium* species and their bioactive constituents has been ignored. This area has recently been reviewed by Selwyn (1979). A number of workers appear to have noted the growth-inhibitory properties of *Penicillium* cultures and Lister used extracts from *Penicillium gluacum* to prevent infection (cited in Selwyn (1979)). The major impediment to a greater understanding (and thus, use) of penicillin appears to have been the instability of the crude preparations. This was also commented on by Fleming in his classic publication (1929).

The use of microbially-derived antibiotics was placed on a firmer scientific footing by the discovery of Fleming that *Penicillium notatum* was inhibitory towards bacterial growth (Fleming, 1929). This species produces a mixture of penicillins (F, G, X and K); however, *P. chrysogenum* was found to be a better source. It was also found that, if phenylacetic acid was added to the culture, it acted as a precursor and penicillin G (benzyl penicillin) was produced. The general strategy of using different precursors to give penicillins with different side chains was carried out with some success; however, in general, severe limitations are imposed (Wolfe *et al.*, 1984). A further strategy was developed by enzymic removal of the side chain at the 6 position leaving 6-amino penicillanic acid which can then be modified chemically (Huber *et al.*, 1972). This has the advantage that certain precursors which cannot be introduced biosynthetically owing to enzyme–substrate incompatibility can be prepared semi-synthetically. Cephalosporins were first isolated from *Cephalosporium acremonium* (Brotzu, 1948). They are naturally less active than penicillins but have been the subject of substantial chemical modification. The general structures of penicillin and cepahlosporin are shown in Fig. 16.

2.6.2 Biosynthesis

The biosynthesis of penicillins and cepahlosporins has been extensively studied and reviewed (Robinson and Gani, 1985). The general pathway is shown in Fig. 17. It would not be possible to review the substantial advances which have been made over the whole biosynthetic pathway in such a short space; therefore only the sulphur introduction and ring formation will be considered. The interested reader is referred

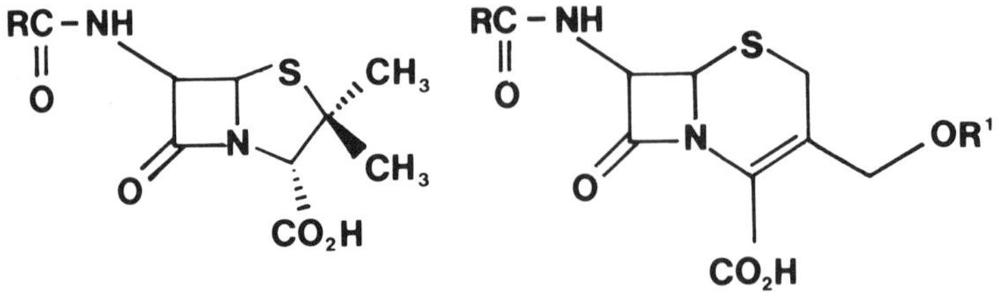

1. Penicillin   **2. Cephalosporin**

Fig. 16 — The structures of penicillin and cepahlosporin.

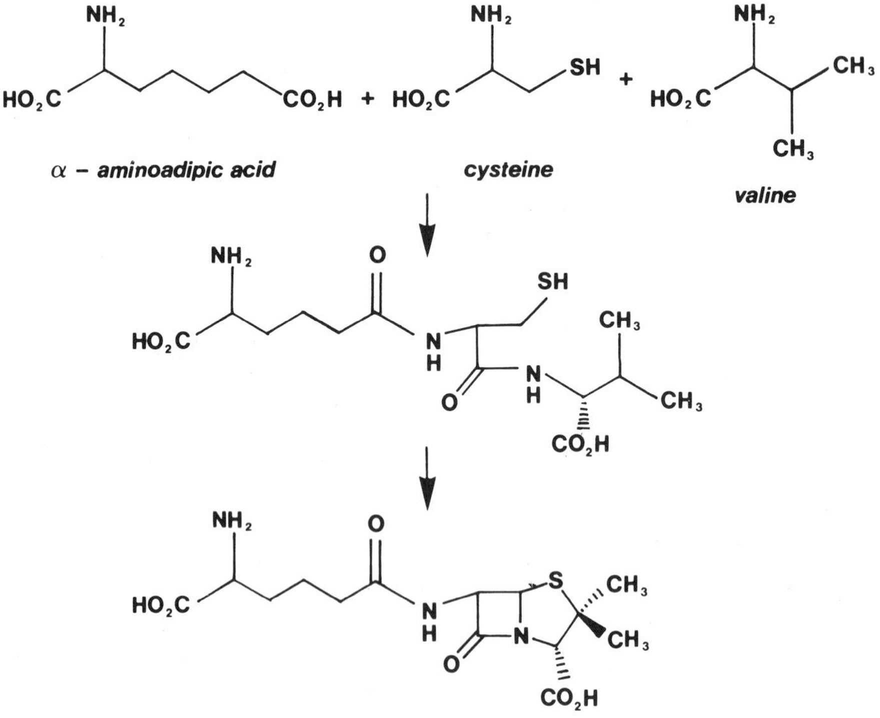

α – aminoadipic acid   cysteine   valine

Fig. 17 — The biosynthesis of penicillins.

to a number of other reviews for a more comprehensive treatment (Aberhart, 1977; Queener and Neuss, 1982).

Incorporation of chiral valine has established that the introduction of sulphur at C-3 of valine occurs with retention of configuration (Baldwin and Wan, 1981; Baxter *et al.*, 1982). Detailed investigation of the ring closure reaction has shown that the S hydrogen is lost from the cysteine residue as shown in Fig. 18 but that no loss of

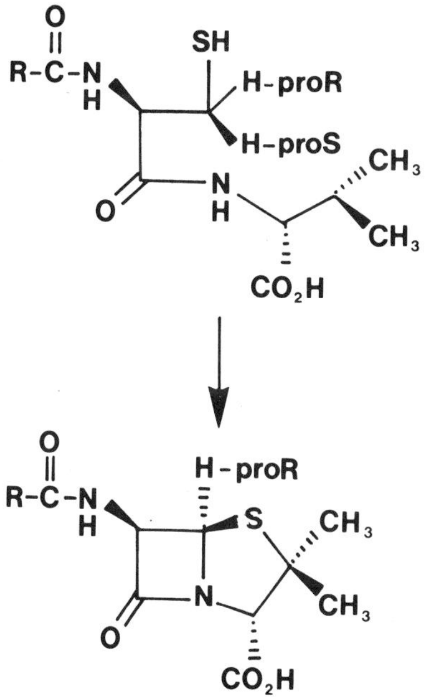

Fig. 18 — The stereochemistry of the ring closure reaction for penicillins.

hydrogen is seen at the adjacent carbon (Morecombe and Young, 1975; Young *et al.*, 1977; Aberhart *et al.*, 1975). Stereospecific closure with retention of configuration is thus implied. The retention of oxygen rules out any possibility of dehydration–rehydration procedures. A proposal that a thiazepine peptide may be involved (Wolfe *et al.*, 1981) (Fig. 19) has been eliminated by synthesis and non-incorporation of this

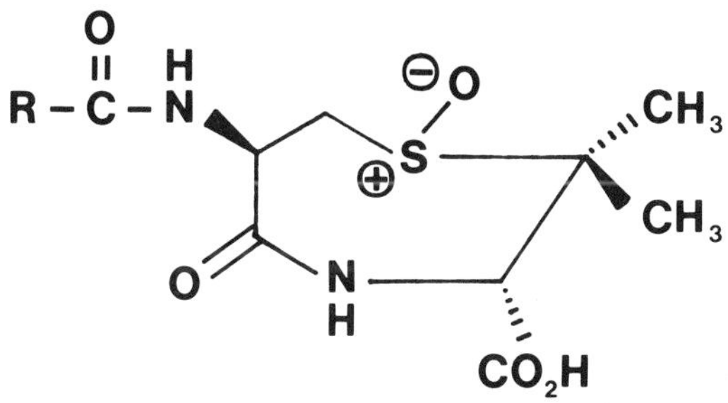

Fig. 19 — A proposed thiazepine peptide intermediate in penicillin biosynthesis.

compound (Bahadur *et al.*, 1981) and related biosynthetic intermediates. Other intermediates have been eliminated; however, it seems likely that the β-lactam ring is formed first during the biosynthesis as an enzyme-bound intermediate. It has been suggested (Baldwin *et al.*, 1984) that the enzyme has strict substrate requirements — particularly with respect to the beginning of the tripeptide (L-α-aminoadipic acid) which must contain a six-carbon chain with a terminal carboxyl group. It has been found recently, however, that phenylacetyl-L-cysteinyl-D-valine can be converted, albeit slowly, to penicillin G by an enzyme from *Acremonium chrysogenum* (Luengo *et al.*, 1986a, b). This penicillin contains a hydrophobic side chain and is unacceptable to the enzyme system which converts penicillins to cephalosporins. It is to be hoped that alteration to this latter enzyme may permit the formation of a hydrophobic cephalosporin which would have distinct advantages in terms of isolation. The fungal expandase–hydroxylase gene from *Cephalosporin acremonium* has recently been cloned and expressed in *E. coli* (Samson *et al.*, 1987).

Similarly, it has been found that a derivative of the tripeptide which is hydroxylated at the site of ring closure (Baxter *et al.*, 1984) or at C-3 (Bahadur *et al.*, 1981) is not incorporated into isopenicillin N. Similarities, therefore, exist between the sulphur ring formation in penicillin, biotin and lipoic acid. In all three cases, hydroxylated derivatives are not involved, nor are unsaturated intermediates and, in the case of biotin and penicillin, the reaction occurs with retention of configuration. The possibility of carbon radical formation followed by trapping has been suggested but remains unproven (Baldwin and Wan, 1979; Baldwin and Davis, 1981; Baldwin *et al.*, 1981).

Isopenicillin N sythetase has been purified (Pang *et al.*, 1984; Kupka *et al.*, 1983; Hillander *et al.*, 1984) and found to require Fe^{2+}, O_2 and ascorbate for maximal activity.

Finally, it should be noted that in the related family of β-lactam antibiotics, the nocardicins which are isolated from several species of the genus *Nocardia*, the mechanism of formation of the β-lactam ring is different from that in penicillins and cepahlosporins. The ring is formed with inversion of configuration (Gani *et al.*, 1983) and the administration of doubly labelled (^{13}C, ^{15}N) precursors suggests that the oxime is generated in the side chain by direct oxidation rather than by transamination (Townsend and Salituro, 1984). The C–N bonds at C-2 and C-5 were both retained.

2.6.3   Modification of antibiotics

As mentioned above, the initial attempts to prepare modified antibiotics arose from feeding suitable precursors to the culture. While such an approach met with some success, the alteration by semisynthetic means (removal of the side chain by amidases and subsequent chemical modification) was adopted enthusiastically and led to substantial developments in synthetic organic chemistry; however, from the thousands of unnatural penicillins which have been synthesized since 1959, only 15 or so are of clinical importance (Wolfe *et al.*, 1984). Recently attempts have been made via site-directed mutagenesis, to modify the enzyme penicillin N synthetase such that suitable precursors can be made and fed to improve the biological spectrum of activity of the product. A detailed explanation of the design and strategy of site-directed mutagenesis is beyond the scope of this chapter, however, and interested readers are referred to other relevant sources (Fersht, 1984).

2.6.4   Non-clinical uses of antibiotics

The principal non-clinical use of antibiotics is in the field of molecular biology. A major method for the introduction of extra chromosomal DNA into cells is by use of small circular pieces of DNA (plasmids) which contain the gene which codes for production of the protein of interest. In many cases, it is difficult to screen individual colonies for production of the protein of interest, i.e. for successful incorporation and production from the plasmid. A number of plasmids have been developed which contain a gene coding for β-lactamase activity. This enzyme breaks down β-lactams, thereby preventing their interference in cell wall synthesis and thus allowing normally susceptible cells to grow. The survival and growth of cells in a culture containing β-lactam antibiotics can, therefore, be used effectively as a primary screen for these cells in the culture which have incorporated the plasmid and are able to make use of it to produce β-lactamase (Old and Primrose, 1985).

2.6.5   Other sulphur-containing antibiotics

A number of other sulphur-containing antibiotics have been isolated from microbial species. The biosynthesis of a number of these compounds has been extensively studied and the mechanism of sulphur incorporation remains an active area of study.

2.7   CONCLUDING REMARKS

While naturally-occurring sulphur compounds are a structurally heterogenous group, it is clear that, in certain respects, they exhibit close chemical, biochemical and biosynthetic similarities. Many of the mechanisms involved in one group of sulphur natural products are applicable to other members of the class. In the case of essential sulphur-containing compounds such as vitamins, amino acids and their derived proteins, the amounts of material produced make detailed investigation difficult. Such problems are more likely to be overcome by stereospecific synthetic strategies aimed towards preparation of precursors combined with mass spectral and gene cloning techniques rather than by the more traditional radioactive precursor feedings adopted hitherto. It is essential, therefore, for bio-organic chemists working in this area to have an appreciation of the advantages which can accrue from the careful use of molecular biology. Such techniques are also likely to become important in the manipulation of secondary products to give novel bioactive compounds. The combination of synthetic chemistry, enzymology and molecular biology is likely to produce many of the most significant advances in the field of naturally-occurring sulphur (and other) compounds in the future.

REFERENCES

Aberhart, D. J. (1977). Biosynthesis of β-lactam antibiotics. *Tetrahedron*, **33**, 1545–1559.

Aberhart, D. J., Lin, L. J. and Chu, J. Y.-R. (1975). Studies on the biosynthesis of β-lactam antibiotics. Part II. *J. Chem. Soc. PPerkin Trans. I*, 2517–2523.

Apontoweil, P. and Berends, W. (1975a). Isolation and initial characterization of glutathione deficient mutants in *Escherichia coli* K12. *Biochim. Biophys. Acta*, **199**, 10–22.

Apontoweil, P. and Berends, W. (1975b). Mapping of gshA, a gene for the biosynthesis of glutathione in *Escherichia coli* K12. *Mol. Gen. Genet.*, **141**, 91–95.

Bahadur, G., Baldwin, J. E., Wan, T., Jung, M., Abraham, ZE. P., Huddleston, J. A. and White, R. L. (1981). On the proposed intermediary of β-hydroxyvaline- and thiazepinone-containing peptides in penicillin biosynthesis. *J. Chem. Soc. Chem. Comm.*, 1146–1147.

Baldwin, J. E. and Davis, A. P. (1981). Penicillin biosynthesis: a model for the oxidative cyclisation of a peptide to a β-lactam. *J. Chem. Soc. Chem. Comm.*, 1219–1221.

Baldwin, J. E. and Wan, T. S. (1979). Penicillin biosynthesis. A model for carbon–sulfur bond formation. *J. Chem. Soc. Chem. Comm.*, 249–250.

Baldwin, J. E. and Wan, T. S. (1981). Penicillin biosynthesis. Retention of configuration at C-3 of valine during its incorporation into the Arnstein tripeptide. *Tetrahedron*, **37**, 1589–1595.

Baldwin, J. E., Beckwith, A. L. J., Davis, A. P., Proctor, G. and Singleton, K. A. (1981). 5-Isothiazoliodinoyl and 5-isoxazoldinosyl radicals. *Tetrahedron*, **37**, 2181–2189.

Baldwin, J. E., Abraham, E. P., Adlington, R. M., Bahadur, G. A., Chakrauarti, B., Domayne-Hayman, B. P., Field, L. D., Flitsch, S. L., Jayarilaka, G. S., Spakouskis, A., Ting, H.-K., Turner, N. J., White, R. L. and Usher, J. J. (1984). Penicillin biosynthesis: active site mapping with aminoadipoylcysteinyl valine variants. *J. Chem. Soc. Chem. Comm.*, 1225–1227.

Baxter, R. L., Scott, A. I. and Fukumura, M. (1982). Retention of configuration at C-3 of (25, 35)-[4-^{13}C]valine in the biosynthesis of δ-(L-α-aminoadipyl)-D-valine, the acyclic precursor of the penicillins. *J. Chem. Soc. Chem. Comm.*, 66–68.

Baxter, R. L., Thomson, G. A. and Scott, A. I. (1984). Synthesis and biological activity of δ-(L-α-aminoadipoyl)-L-cysteinyl-N-hydroxy-D-valine: a proposed intermediate in the biosynthesis of the penicillins. *J. Chem Soc. Chem. Comm.*, 32–34.

Bowen, C. E., Rauscher, E. and Ingraham, L. L. (1968). The basicity of biotin. *Arch Biochem. Biophys.*, **125**, 865–872.

Bränden, C.-I., Jornuall, H., Eklund, H. and Furusren, B. (1975). Alcohol dehydrogenases. *Enzymes*, **11**, 104–186.

Breslow, R. (1958). On the mechanism of thiamine action IV. Evidence from studies on model systems. *J. Amer. Chem. Soc.*, **80**, 3719–3723.

Brotzu, G. (1948). *Labori dell'Instituto d'Igiene de Cagliari*, p. 1.

Brown, G. M. and Williamson, J. M. (1982). Biosynthesis of riboflavin, folic acid, thiamine and pantothenic acid. *Adv. Enzymol.*, **53**, 345–381.

Bullock, M. W., Brockman, J. A., Patterson, E. L., Pierce, J. V., von Saltza, M. H., Sanders, F. and Stockstad, E. L. R. (1954). Synthesis in the thioctic acid series. *J. Amer. Chem. Soc.*, **76**, 1828–1832.

Butt, T. R. and Ecker, D. J. (1987). Yeast metallothionein and applications in biotechnology. *Microbiol. Rev.*, **51**, 351–364.

Carbon, J. A., Hung, L. and Jones, D. S. (1965). The reversible oxidative inactivation of specific transfer RNA species. *Proc. Natl. Acad. Sci. USA*, **53**, 979–986.

Cooper, A. J. C. (1983). The biochemistry of sulfur-containing amino acids. *Ann. Rev. Biochem.*, **52**, 187–222.

Cooper, A. J. C., Habel, M. T. and Meister, A. (1982). On the chemistry and biochemistry of 3-mercaptopyruvic acid, the α-keto analogue of cysteine. *J. Biol. Chem.*, **257**, 816–826.

Crawhall, J. C. and Segal, S. (1967). The intracellular ratio of cysteine and cystine in various tissues. *Biochem. J.*, **105**, 891–896.

Datko, A. H., Mudd, S. H. and Giovanelli, J. (1977). Homocysteine biosynthesis in green plants. *J. Biol. Chem.*, **252**, 3436–3445.

DeMoll, E. and Shive, W. (1985). Determination of the metabolic origin of the sulfur atom in thiamine of *Escherichia coli* by mass spectrometry. *Biochem. Biophys. Res. Comm.*, **132**, 217–222.

de Rey-Pailhade, J. (1888a). Nouvelles recherchesa physialogiques sur la substance organique hydrogénant le soufre à froid. *Comptes Rendus*, **107**, 43–44.

de Rey-Pailhade, J. (1888b). Sur un corps d'origine organique hydrogénant le soufre à froid. *Comptes Rendus*, **106**, 1683–1684.

De Titta, G. T., Edmonds, J. W., Stallings, W. and Donohue, J. (1976). Molecular structure of biotin. Results of two independent crystal structure investigations. *J. Amer. Chem. Soc.*, **98**, 1920–1926.

Dixon, J. C. and Benevenga, N. J. (1980). The decarboxylation of α-keto-γ-methyl butyrate in rat liver mitochondria. *Biochem. Biophys. Res. Comm.* **97**, 939–946.

Douglas, K. T. (1987). Mechanism of action of glutathione dependent enzymes. *Adv. Enzymol.*, **59**, 103–167.

Eisenberg, M. A. (1973). Biotin: biogenesis, transport and their regulation. *Adv. Enzymol.*, **38**, 317–372.

Eklund, H., Plapp, B. V., Samama, J. P. and Branden, C.-I. (1982). Binding of substrate in a ternary complex of horse liver alcohol dehydrogenase. *J. Biol. C hem.*, **257**, 14349–14358.

Estramareix, B., Gaudry, D. and Therisod, M. (1977). Biosynthèse du thiazole de la thiamine chez *Escherichia coli*. *Biochimie*, **59**, 857–859.

Fersht, A. (1984). *Enzyme Structure and mechanism*, 2nd edn., Freeman, New York, pp. 382–387.

Flavin, M. and Slaughter, C. (1971). In H. Tabor and C. W. Tabor (eds.), *Methods in Enzymology*, Vol. XVII, Part B, Academic press, New York, pp. 422–439.

Fogel, S. and Welch, J. W. (1982) Tandem gene amplification mediates copper resistance in yeast. *Proc. Natl. Acad. Sci. USA*, **79**, 5342–5346.

Fleming, A. (1929). On the antibacterial action of cultures of a penicillium with special reference to their use in the isolation of *B. influenzae*. *Br. J. Exp. Path.*, **10**, 226–236.

Frappier, F., Guillerm, G., Salib, A., G. and Marquet, A. (1979). On the mechanism of conversion of dethiobiotin to biotin in *E. coli*. Discussion of the occurrence of an intermediate hydroxylation. *Biochem. Biophys. Res. Comm.*, **91**, 521–527.

Frappier, F., Jouany, M., Marquet, A., Olesker, A. and Tabbet, J.-C. (1982). On the mechanism of the conversion of dethiobiotin to biotin in *E. coli*. Studies with deuterated precursors using tandem mass spectroscopic (MS–MS) techniques. *J. Org. Chem.*, **47**, 2257–2261.

Fry, D. C., Fox, T. L., Lane, M. D. and Mildvan, A. S. (1985a). NMR studies of the exchange of the amide protons of d-biotin and its derivatives. *Ann. NY Acad. Sci.*, **447**, 140–151.

Fry, D. C., Fox, T. L., Lane, M. D. and Mildvan, A. S. (1985b). Exchange characteristics of the amide protons of d-biotin and derivatives. *J. Amer. Chem. Soc.*, 107, 7659–7665.

Fuchs, J. A. and Warner, H. R. (1975). Isolation of an *Escherichia coli* mutant deficient in glutathione synthesis. *J. Bacteriol.*, **124**, 140–148.

Gani, D., Young, D. W., Carr, D. M., Pyser, J. P. and Sadler, I. H. (1983). Synthesis of a monocyclic β-lactam stereospecifically labelled at C-4. *J. Chem. Soc. Perkin Trans. I*, 2811–2814.

Giovanelli, J. and Mudd, S. H. (1967). Synthesis of homocysteine and cysteine by enzyme extracts of spinach. *Biochem. Biophys. Res. Comm.*, 27, 150–156.

Giovanelli, J. and Mudd, S. H. (1971). Transsulfuration in higher plants. Partial purification and properties of β-cystathione of spinach. *Biochim. Biophys. Acta*, **227**, 654–670.

Giovanelli, J., Mudd, S. H. and Datko, A. H. (1978). Homocysteine biosynthesis in green plants. *J. Biol. Chem.*, **253**, 5665–5677.

Grue-Sørensen, G., White, R. L. and Spenser, I. D. (1986). Thiamin biosynthesis in *Saccharomyces cerevisae*: origin of the pyrimidine unit. *J. Amer. Chem. Soc.*, **108**, 146–158.

Guchait, R. B., Polakis, S., E., Hollis, D., Fensezau, C. and Lane, M. D. (1974). Acetyl coenzyme A carboxylase system of *Escherichia coli*. *J. Biol. Chem.*, **249**, 6646–6656.

Guillerm, F., Frappier, F., Gaudry, M. and Marquet, A. (1977). On the mechanism of conversion of dethiobiotin to biotin in *Escherichia coli*. *Biochemie*, **59**, 119.

Guirard, B. M., Snell, E. E. and Williams, R. J. (1946). The nutritional role of acetate for lactic acid bacteria. II. Fractionation of extracts of natural materials. *Arch. Biochem. Biophys.*, **9**, 381–388.

Haslam, E. (1985). *Metabolites and Metabolism — a Commentary on Secondary Metabolism*, Clarendon, Oxford.

Hillander, I. J., Shen, Y.-G., Heim, J., Demain, A. C. and Wolfe, S. (1984). A pure enzyme catalysing penicillin biosynthesis. *Science*, **224**, 610–612.

Hitchcock, C. H. and Walker, J. (1961). Experiments on the biosynthesis of thiamine. *Biochem. J.*, **80**, 137–148.

Hopkins, F. G. (1921). On an autoxidisable constituent of the cell. *Biochem. J.*, **15**, 286–305.

Huber, F. M., Chauvette, R. R. and Jackson, B. G. (1972). In E. H. Flynn (ed.), *Cephalosporins and Penicillins. Chemistry and Biology*, Academic Press, New York, p. 27.

Itakura, K., Hirose, T., Crea, R., Riggs, A. D., Heyneker, H. L., Bolivar, F. and Boyer, H. W. (1977). Expression in *E. coli* of a chemically synthesised gene for the hormone somatostatin. *Science*, **198**, 1056–1063.

Kagi, J. H. R. and Nordberg, N. (eds.) (1979). *Metallothionein*, Birkhauser, Basel.

Kaplan, M. M. and Flavin, M. (1965). Enzymatic synthesis of L-cystathionine from the succinic ester of L-homoserine. *Biochem. Biophys. Acta*, **104**, 390–396.

Karin, M. R., Najarain, A., Haslinger, P., Valenzuela, P., Welch, J. and Fogel, S. (1984). Primary structure and transcription of an amplified genetic locus: the CUP 1 locus of yeast. *Proc. Natl. Acad. Sci. USA*, **81**, 337–341.

Kisselev, L. L. and Favorova, O. O. (1974). Aminoacyl-t-RNA syntheses: some recent results and achievements. *Adv. Enzymol.*, **40**, 141–238.

Kogl, F. and Ton nis, B. (1936). Uber das Bios-Problem. Darstellung von krystalliscertem Biotin aus Eigelb. *Z. Physiol. Chem.*, **242**, 43.

Kozluk, T. and Spencer, I. D. (1987). ^{13}C NMR spectroscopy as a biosynthetic probe: the biosynthesis of purines in yeast. *J. Amer. Chem. Soc.*, **109**, 4698–4702.

Kredich, N. M., Becker, M. A. and Tomkin, G. H. (1979). Purification and characterisation of cysteine synthetase, a bifunctional protein complex, from *Salmonella typhimurium*. *J. Biol. Chem.*, **244**, 2428–2439.

Kreis, W. (1979). Tumor therapy by deprivation of L-methionine: rationale and results. *Cancer Treat. Rep.*, **63**, 1069–1072.

Kreis, W., Baker, A., Ryan, V. and Bertasso, A. (1980). Effect of nutritional and enzymatic methionine deprivation upon human normal and malignant cells in tissue culture. *Cancer Res.*, **40**, 643–641.

Kupka, J., Shen, Y.-Q., Wolfe, S. and Demain, A. C . (1983). Studies on the ring-cyclization and ring-expansion enzymes of β-lactam biosynthesis in *Cepahlosporium acremonium*. *Can. J. Microbiol.*, **29**, 488.

Lane, M. D., Young, D. L. and Lynen, F. (1964). The enzymatic synthesis of holotranscarboxylase from apotranscarboxylase and (+)-biotin. *J. Biol. Chem.*, **239**, 2858–2864.

Lazzarini, R. A. and Atkinson, D. E. (1961). A triphosphopyridine nucleoside-specific nitrite reductase from *Eschericia coli*. *J. Biol. Chem.*, **236**, 3330–3336.

Leder, I. G. (1970). In D. B. McCormick and L. D. Wright (eds.), *Methods in Enzymology*, Vol. 18A, Academic Prerss, New York, p. 207.

Leder, I. G. (1975). Thiamine biosynthesis and function. In D. M. Greenberg (ed.), *Metabolic Pathways*, Vol. 7, Academic Press, New York, pp. 57–85.

Lipmann, F., Kaplan, N. O., Novelli, G. D., Tuttle, L. C. and Guirard, B. M. (1947). Coenzyme for acetylation, a panthothenic acid derivative. *J. Biol. Chem.*, **167**, 869–870.

Lohmann, K. and Schuster, P. (1937). Uber die Co-carboxylase. *Naturwissenschaften*, **25**, 26–27.

Luengo, J. M., Alemany, M. R. , Salto, F., Ramos, F., Lopez-Nieto, M. J. and Martin, J. F. (1986a). Direct enzymatic synthesis of penicillin G using cyclases of *Penicillium chrysogenum* and *Acremonium chrysogenum*. *Biotechnololgy*, **4**, 44–47.

Luengo, J. M., Imso, J. L. and Lopez-Nieto, M. J. (1986b). Direct enzymatic synthesis of natural penicillins using phenylacetyl CoA. *J. Antibiot.*, **29**, 1754–1759.

Manicoll, P. K., Datko, A. H., Givanelli, J. and Mudd, S. H. (1981). Homocysteine biosynthesis in green plants: physiological importance of the transsulfuration pathway in *Lemna pancicostata*. *Plant Physiol.*, **68**, 619–625.

Mapson, L. W. E. (1969) Biogenesis of ethylene. *Biol. Rev.*, **44**, 155–187.

McCluskey, T. J., Scarf, A. R. and Anderson, J. W. (1986). Enzyme catalysed α,β-elimination of selenocystathionine and selenocystine and their sulfur analogues by plant extracts. *Phytochemistry*, **25**, 2063–2068.

Meister, A. and Anderson, M. E. (1983). Glutathione. *Ann. Rev. Biochem.*, **52**, 711–760.

Melville, D. B., Moyer, A. W., Hofmann, K. and duVigneaud, V. (1942). The structure of biotin: the formation of thiophene valeric acid from biotin. *J. Biol. Chem.*, **146**, 487–493.

Mildvan, A. S. and Fry, D. C. (1987). NMR studies of the mechanism of enzyme action. *Adv. Enzymol.*, 241–313.

Mildvan, A. S., Scrutton, M. C. and Utter, M. F. (1966). Pyruvate carboxylase. *J. Biol. Chem.*, **241**, 3488–34398.

Mislow, K. and Meluch, W. C. (1956). The stereochemistry of α-lipoic acid. *J. Amer. Chem. Soc.*, **78**, 5920–5923.

Morecombe, D. J. and Young, D. W. (1975). Synthesis of chirally-labelled cysteines and the steric origin of C(5) in penicillin biosynthesis. *J. Chem. Soc. Chem. Comm.*, 198–199.

Murakoshi, I. and Kaneko, M. (1985). Purification and properties of cysteine synthase from *Spinicia oleracea. Phytochemistry*, **24**, 1907–1912.

Murakoshi, I., Kaneko, M., Koide, C. and Ikegami, F. (1986). Enzymatic synthesis of the neuroexcitatory amino acid quisqualic acid by cysteine synthase. *Phytochemistry*, **25**, 2759–2763.

Naiki, N. (1965). Some properties of sulfite reductase from yeast. *Plant Cell Physiol. (Japan)*, **6**, 179–194.

Newell, P. C. and Tucker, R. G. (1967). New pyrimidine pathway involved in the biosynthesis of the pyrimidine of thiamine. *Nature* **215**, 1384–1385.

Newell, P. C. and Tucker, R. G. (1968a). Precursors of the pyrimidine moiety of thiamine. *Biochem. J.*, **106**, 271–277.

Newell, P. C. and Tucker, R. G. (1986b). Biosynthesis of the pyrimidine moiety of thiamine. *Biochem. J.*, 106, 279–287.

Old, R. W. and Primrose, S. B. (1985). *Principles of Gene Manipulation*, 3rd edn., Blackwell, Oxford.

Ozaki, H. (1986). Reinvestigation of compound X, a suspected biotin intermediate. *J. Nutr. Sci. Vitaminol.*, **32**, 279–286.

Pang, C.-P., Chakrauarti, B., Adlington, R. H., Ring, H.-H., White, R. L., Jayatilake, E. S., Baldwin, J. E. and Abraham, E. P. (1984). Purification of isopenicillin N. synthetase. *Biochem. J.*, **222**, 789–795.

Parry, R. J. (1977). Biosynthesis of lipoic acid I. Incorporation of specifically tritiated octanoic acid into lipoic acid. *J. Amer. Chem. Soc.*, **99**, 6464–6466.

Parry, R. J. (1983). Biosynthesis of some sulfur-containing natural products. Investigations of the mechanism of carbon-sulfur bond formation. *Tetrahedron*, **39**, 1215–1238.

Parry, R. J. and Naidu, M. U. (1980). Biotin biosynthesis. Incorporation of 5(RS)-^{3}H-dethiobiotin into biotin. *Tetrahedron. Lett.*, 4783–4786.

Parry, R. J. and Trainor, D. A. (1978). Biosynthesis of lipoic acid. 2. Stereochemistry of sulfur introduction of C-6 of octanoic acid. *J. Amer. Chem. Soc.*, **100**, 5243–5244.

Pearson, B. M., Mackenzie, D. A. and Keenan, M. H. J. (1986). Production of biotin by yeasts. *Lett. Appl. Microbiol.*, **2.**, 25–28.

Premakumar, R., Winge, D. R., Wiley, R. D. and Rajogpalan, K. V. (1975). Copperchelatin: isolation from various eukaryotic sources. *Arch. Biochem. Biophys.*, **170**, 278–288.

Prinz, R. and Weser, U. (1975). Naturally occurring Cu-thionein in *Saccharomyces cervisae. J. Physiol. Chem.*, **356**, 767–776.

Queener, S. W. and Neuss, N. (1982). In R. B. Morin and M. Forman (eds.), *The Chemistry and Biology of β-lactam Antibiotics*, Vol. 3, Academic Press, New York, p. 1.

Reed, C. J. (1966). Chemistry and function of lipoic acid. In M. Florkin and E. Stotz (eds.), *Comprehensive Biochemistry*, Vol. 14, Elsevier, New York, pp. 99–127.

Reed, L. J., DeBusk, B. G., Gunsalus, I. C. and Hornbergerir, C. S. (1951). Crystalline α-lipoic acid: a catalytic agent associated with pyruvate dehydrogenase. *Science*, **114**, 93–94.

Robinson, J. A. and Gani, D. (1985). Enzymology in biosynthesis: mechanistic and stereochemical studies of β-lactam biosynthesis and the shikimate pathway. *Nat. Prod. Rep.*, **2**, 293–320.

Salib, A. G., Frappier, F., Fuiller, M. G. and Marquet, A. (1979). On the mechanism of the conversion of dethiobiotin to biotin in *E. coli* III. *Biochem. Biophys.*, *Res. Comm.*, **88**, 312–319.

Samson, S. M., Dotzlaf, J. E., Elislz, M. L., Becker, G. W., Van Frank, R. M., Veal, L. E., Yeh, W.-K., Miller, J. R., Queener, S. W. and Ingolia, T. D. (1987). Cloning and expression of the fungal expandase/hydroxylase gene involved in cepahlosporin biosynthesis. *Biotechnology*, **5**, 1207–1214.

Schlossman, K. and Lynen, F. (1957). Biosynthese deslysteins aus Serin und Schwefelwasserstoff. *Biochem. Z.*, **328**, 591–594.

Selwyn, S. J. (1979). Pioneer work on the "penicillin phenomenon", 1870–1876. *Antimicrob. Chemother.*, **5**, 249.

Shimazone, N., Mano, Y., Tanaka, R. and Kaziro, Y. (1959). Mechanism of transpyrophosphorylation with thiaminokinase. *J. Biochem. (Tokyo)*, **46**, 959–961.

Singer, T. P. (1975). In D. M. Greenberg, (ed.), *Metabolic Pathways*, Vol. 7, *Metabolism of Sulfur Compounds*, Academic press, New York, pp. 535–546.

Steele, R. D., Barber, T. A., Lalich, J. and Benevenga, N. J. (1979). Effects of dietary 3-methylthiopropionate on metabolism, growth and hematopoiesis in the rat. *J. Nutr.*, **104**, 1739–1751.

Tamura, G., Asada, K. and Bandurski, R. S. (1967). Linkage sulphate reduction to pyridine nucleotide. *Plant Physiol.*, **42**, 536.

Tazuya, K., Yamamoto, M., Hayashiji, M., Yamada, K. and Kumaoka, H. (1986). Biosynthesis of thiamine. Precursor of C-5, C-6 and hydroxymethyl carbon atoms of the pyrimidine moiety in a eukaryote. *Biochem. Int.*, **12**, 661–668.

Therisod, M., Gaudry, D. and Estramareix, B. (1978). The biosynthesis of thiamine: isolation of a new thiazolic metabolite of tyrosine. *Nouv. J. Chim.*, **2**, 119–121.

Townsend, C. A. and Salituro, G. M. (1984). Fate of [^{15}N]-(p-hydroxyphenyl) glycine in Nocardicin A biosynthesis. *J. Chem. Soc. Chem. Comm.*, 1631–1632.

Trainor, D. A., Parry, R. J. and Gitterman, A. (1980). Biotin biosynthesis 2. Stereochemistry of sulfur introduction at C-4 of dethiobiotin. *J. Amer. Chem. Soc.*, **102**, 1467–1468.

Wakil, S. J., Titchener, E. B. and Gibson, D. M. (1958). Evidence for the participation of biotin in the enzymic synthesis of fatty acids. *Biochem. Biophys. Acta*, **29**, 225–226.

White, R. H. (1980a) Stable isotope studies on the biosynthesis of lipoic acid on *Escherichia coli*. *Biochemistry*, **19**, 15–19.

White, R. H. (1980b). Biosynthesis of lipoic acid: extent of incorporation of deuterated hydroxy and thio-octanoic acids into lipoic acid. *J. Amer. Chem. Soc.*, **102**, 6605–6607.

White, R. H. (1980c). Stoichiometry and stereochemistry of deterium incorporated into fatty acids by cells of *Escherichia coli* grown on [methyl-^{2}H$_3$]-acetate. *Biochemistry*, **19**, 9–15.

White, R. H. and Rudolph, F. B. (1978). The origin of the nitrogen atom in the thiazole ring of thiamine in *Escherichia coli*. *Biochem. Biophys. Acta*, **542**, 340–347.

White, R. L. and Spencer, I. D. (1982). Thiamin biosynthesis in yeast. Origin of the five carbon unit of the thiazole moiety. *J. Amer. Chem. Soc.*, **104**, 4934–4943.

Wolfe, S., Bowers, R. J., Hasan, S. K. and Kazmaier, P. M. (1981). Synthesis and conformation of α-aminoadipyl and glycyl-α-aminoadipyl thiazepine sulfoxides. *Can. J. Chem.*, **59**, 406–421.

Wolfe, S., Demain, A. L., Jensen, S. E. and Westlake, D. W. S. (1984). Enzymatic approach to the syntheses of unnatural β-lactams. *Science*, **226**, 1386–1392.

Wood, H. E. and Barden, R. E. (1977). Biotin enzymes. *Ann. Rev. Biochem.*, **46**, 385–412.

Yamada, K., Yamamoto, M., Hayashiji, M., Tazuya, K. and Kumaoka, H. (1984). Biosynthesis of thiamin. The precursor of the five carbon unit of the thiazole moiety. *Biochem. Int.*, **10**, 689–694.

Young, D. W. (1986) The biosynthesis of the vitamins thiamine, riboflavin and folic acid. *Natl. Prod. Rep.*, **3**, 395–419.

Young, D. W., Morecombe, D. J. and Sen, P. K. (1977). The stereochemistry of β-lactam formation in penicillin biosynthesis. *Eur. J. Biochem.*, **75**, 133–147.

3

Agricultural importance of sulphur xenobiotics

Gerald T. Brooks
AFRC, University of Sussex, Brighton BN1 9RQ, UK[†]

SUMMARY

1. Some 30% of the chemicals currently used as insecticides, acaricides, nematicides, herbicides, plant growth regulators or fungicides contain sulphur.
2. Sulphur has a major role as a versatile constituent of propesticides. Thus, the oxidation of sulphur *in vivo* may occur with its retention and result in favourable changes in the physical properties of molecules, as in the improved systemic activity of thioether insecticides. Oxidative bioactivation may also generate reactive molecules by the elimination of sulphur, as in the case of certain organophosphorus anticholinesterase insecticides, or activate constituent molecules by generating readily displaceable, sulphur-containing leaving groups, as with some herbicides. Furthermore, the cleavage of appropriate sulphur xenobiotics by hydrolysis or reduction may generate thiophenols or other biologically active fragments. However, these reactions may also facilitate rapid elimination of sulphur xenobiotics from the environment.
3. The direct toxic action of sulphur xenobiotics through interaction of their sulphur-containing functional groups with specific molecular target sites shown to be critical for survival is not well documented.

3.1 INTRODUCTION

Elemental sulphur has long been used by itself as a non-systemic direct and protective fungicide, insecticide and acaricide. It is also used in the form of lime sulphur (calcium polysulphide; CaS_x), which decomposes in the presence of carbon

† *Present address:* Department of Biochemistry and Physiology, University of Reading, Whiteknights, Reading RG6 2AJ, UK.

dioxide and acids to give elemental sulphur and hydrogen sulphide. The well-known 'Bordeaux mixture' is a mix of calcium hydroxide and copper sulphate which was introduced in 1885 as a protective fungicide for applications on foliage. Apart from this, the value of sulphur as a component of synthetic chemicals used in crop protection and for public health purposes is revealed by an examination of the 8th edition of *The Pesticide Manual* (Worthing and Walker, 1987) which lists in the 'current use' section some 104 sulphur-containing compounds with insecticidal, acaricidal or nematicidal properties, 52 herbicides–plant growth regulators (PGR) and 34 fungicides. This represents about 30% of the total entry; the predominance of insecticidal compounds reflects the large number and structural variety of organophosphorus insecticides containing sulphur.

In the following account, sulphur compounds are considered in subsections on insecticides, herbicides or fungicides. However, some compounds have uses in more than one of these areas and may also find applications in public health as well as agricultural pest control. Summaries of chemical properties, uses, toxicology, formulations and analysis are to be found in *The Pesticide Manual*. The discussion in this chapter is based as far as possible on recently reported research. For work prior to 1983, the reader's attention is drawn when appropriate to reviews of works relating to pesticides in general (Rosen *et al.*, 1981; Cole, 1983; Muecke, 1983; Brooks, 1984), insecticides (Brooks, 1974, 1979; Kuhr and Dorough, 1976; Knowles, 1982), herbicides (Dodge, 1983; Esser, 1986) and fungicides (Langcake *et al.*, 1983; Vonk, 1983; Somerville, 1986).

3.2 INSECTICIDES, ACARICIDES, NEMATICIDES

This is the second largest pesticide group in economic importance, after herbicides, accounting for about 33% of worldwide pesticide sales in 1983 (Somerville, 1986).

3.2.1 Fumigants

Apart from elemental sulphur, carbon disulphide, which was first used as an insecticide in 1854, is probably the sulphur xenobiotic best known in pest control. The low boiling point (46.3°C) and action as a respiratory poison have led to its wide use as a fumigant for nursery stock and in the treatment of soil to control insects and nematodes. Fumigants have an important role in the protection of stored grain from insect pests, especially in tropical and subtropical countries. Their ability, when used correctly, to exterminate all stages of insect pests in bulked or baggaged stored products has been used either alone, or frequently in mixtures with volatile liquid chlorohydrocarbons such as ethylene dichloride, ethylene dibromide and carbon tetrachloride. Simple fumigants can be applied in the form of chemical precursors, a principle widely used in insect control. Thus the soil fumigant dazomet (**1**) affords CS_2 by acid hydrolysis but in soil mainly breaks down into methyl isothiocyanate, also used as a soil fumigant both directly and in the form of another precursor, methyldithiocarbamic acid ($CH_3NHCS \cdot SH$).

(1)

3.2.2   Organochlorine insecticides and pyrethroids

Sulphur has a prominent position in the development of the modern organic insecticides, as a component of various dyes and water-soluble moth-proofing agents investigated in Switzerland and Germany before World War II. Work on sulphur-containing stomach poisons for the clothes moth led to a search for more lipophilic chemicals of possible use as contact insecticides in plant protection and eventually to the discovery of DDT.

Many of the DDT forerunners (2; R= $-SO_2NHCO-$, $-O \cdot SO_2-$, $-S-$, $-SO-$, $-SO_2-$; R_1, R_2 = Cl) were active insecticidal stomach poisons and a number of outstanding acaricides and ovicides were subsequently based on this type of structure (Brooks, 1974). Chlorfenson, (2; R_1, R_2 = Cl; R = $-O \cdot SO_2-$), tetrasul (4-chlorophenyl 2,4,5-trichlorophenyl sulphide) and its sulphone (tetradifon), and fenson (4-chlorophenyl benzenesulphonate) are still used as acaricides today. In the course of investigations on resistance, N-(di-n-butyl)-4-chloro benzenesulphonamide (3; WARF antiresistant) was found to potentiate DDT (2; R_1, R_2 = Cl; R = $CHCCl_3$) against DDT-resistant houseflies by inhibiting the enzyme DDT-dehydrochlorinase (DDT-ase). DDT-ase converts DDT into DDE (2; R_1, R_2 = Cl; R = $C=CCl_2$), a common, although not the only, cause of DDT resistance in insects. This early example of insecticide synergism by a sulphur xenobiotic set the scene for many future studies of insecticide metabolism.

(2)   (3)

Some of the acaricides mentioned obviously represent successive stages in the probable oxidative metabolism of the sulphide link in bis(4-chlorophenyl)sulphide (2; R_1, R_2 = Cl; R=S). Subsequent experience has shown that the successive conversion of the sulphide moiety into sulphoxide and sulphone is a valuable attribute and nowadays a major reason for its incorporation into pesticide molecules. This metabolic conversion can alter the toxicokinetics of pesticides and the products may represent detoxication, or be equally toxic or more toxic than the parent molecules (bioactivation). In the case of DDT, which has been well investigated in an attempt to improve its environmental acceptability, replacement of each of the aromatic 4-chlorine substituents by CH_3S- gives increased biodegradability resulting from attack on these moieties by mixed function oxidases (MFO). The intrinsic insect toxicity of methiochlor (2; R_1, R_2 = $-SCH_3$; R=$CHCCl_3$), as measured in the presence of an MFO inhibitor such as piperonyl butoxide (PBO) *in vivo*, is useful, although somewhat lower than that of DDT. The electron-donating thiomethyl groups reduce the rate of enzymic dehydrochlorination, which is desirable, but sulphoxidation, although enhancing biodegradability, has the opposite effect. This illustrates the problems inherent in design from metabolism studies. A better combination is 4-thiomethyl with 4'-ethoxy; the resulting molecule approaches DDT in toxicity to some diptera and may exceed it when applied with PBO, especially toward resistant insects having the DDT-ase mechanism. Also, such molecules normally have a lower mammalian toxicity than DDT and a lower tendency to accumulate in adipose tissue because of their susceptibility to oxidative conversions. These compounds are experimental insecticides but are included here as an illustration of important principles.

A less complicated example is the cyclodiene insecticide endosulfan (**4**), a cyclic sulphite ester. In this case, the competing metabolic routes in both insects and vertebrates are hydrolysis to the parent diol and further oxidation to the toxic sulphate (**5**). A moderate synergism of the insect toxicity of endosulfan by PBO suggests that endosulfan has a greater intrinsic toxicity than the sulphate. Although the practical use of cyclodiene insecticides has declined, particularly in developed countries, endosulfan is a biodegradable compound which still has a number of important uses. e.g. in tsetse fly control and for a number of agricultural purposes, alone or in combination with other insecticides such as the synthetic pyrethroids. It is likely to become the major agricultural insecticide in India (Murkerjee, 1982).

Of the many modifications of the 2,2-dimethyl-3-isobutenyl side chain of synthetic pyrethroids which have been described, that seen in kadethrin is of particular interest in the present context. While the methoxycarbonyl moiety found in natural pyrethrin II confers moderate insect toxicity in that compound and fair knock-down efficiency in the 5-benzyl-3-furylmethyl alcohol derivative (**6**), conversion of the ester substituent into the cyclic thiolactone moiety, as in kadethrin (**7**; RU15525), produces an extremely potent knock-down agent for diptera, although this is at the expense of ultimate toxicity. Such compounds have valuable sanitation uses.

3.2.3   Organophosphorus insecticides
The almost unlimited possibilities for structural variations which can be used to introduce species-selective toxicity or to provide new compounds effective against organophosphorus (OP) resistant insects have made this class favoured as replacements for the persistent organochlorine (OC) insecticides. Well-known phosphorothionates such as parathion (**8**) and malathion (**9**) are pro-insecticides, which for toxicity require oxidative conversion into the corresponding phosphates, paraoxon

and malaoxon (P=S→P=O; desulphuration) that are the active acetylcholinesterase (AChE) inhibitors. This conversion occurs universally in living organisms and toxicity is determined by the bioavailability of the oxons, which in turn depends on the balance between oxidative bioactivation and the enzymic detoxication of both parent phosphorothionate and product oxon. Paraoxon and malaoxon, produced by the MFO of living tissues and possibly also by peroxidases in plants, are about 1.5×10^4 and 4×10^3 more potent than the corresponding parents as AChE inhibitors. The practical use of the pro-insecticidal phosphorothionates (P=S) instead of the corresponding phosphates (P=O) provides an opportunity, which often favours mammals ('opportunity factor'), for the organism to metabolize the pro-insecticide by routes leading to detoxication rather than to the toxic oxon. This is reflected in the relative LD_{50}s (Table 1).

Table 1 — Developments in structural modifications to improve the mammalian safety of organophosphorus and carbamate anticholinesterase insecticides

Compound	Acute oral LD_{50} (mg/kg)	
	Rat	Mouse
Paraoxon (P=O)	3.5	
Parathion (**8**)	13	25
(4-NO$_2$PhO)P(O)(OCH$_3$)$_2$ (methyl paraoxon)		15–20
Methyl parathion (P=S)	14	99
(3-CH$_3$,4-NO$_2$PhO)P(O)(OCH$_3$)$_2$		150–200
(3-CH$_3$,4-NO$_2$PhO)P(S)(OCH$_3$)$_2$ (**21**)	800	1500
Profenofos (=O; **13**)	358	
Prothiofos (**12**; P=S analogue)	966	
Malaoxon (**9**; P=O analogue)	158	
Malathion (**9**)	10 000 (when pure)	1000
Isomalathion (**20**)	120	
Methamidophos (**14**)	30	30
Acephate (**15**)	945	361
Carbofuran (**30**)	8–14	11
Carbosulfan (**31**)	209	129
Furathiocarb (**32**)	106	130
Benfuracarb (**35**)	138	175
DCDM-S-carbofuran (**45**)		152
Methomyl (**29**)	25	16
Thiodicarb (**33**)	431	
U-56295 (**34**)	>8000	

Values from Drabek and Neumann (1985), Eya and Fukuto (1986), Fahmy *et al.* (1978), Metcalf and Metcalf (1973), Ryan and Fukuto (1985), Umetsu (1986) and Worthing and Walker (1987).

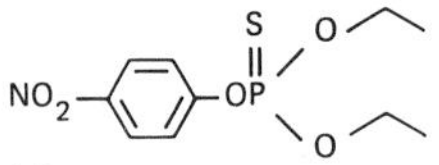

(8) 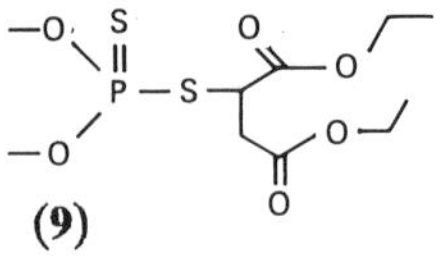   (9)

Desulphuration is believed to involve the generation of a cyclic intermediate with oxygen at the P=S bond (see Chapter 1, this volume, Part B), which then gives the oxon by rearrangement and other observed products by hydrolysis (Scheme 1).

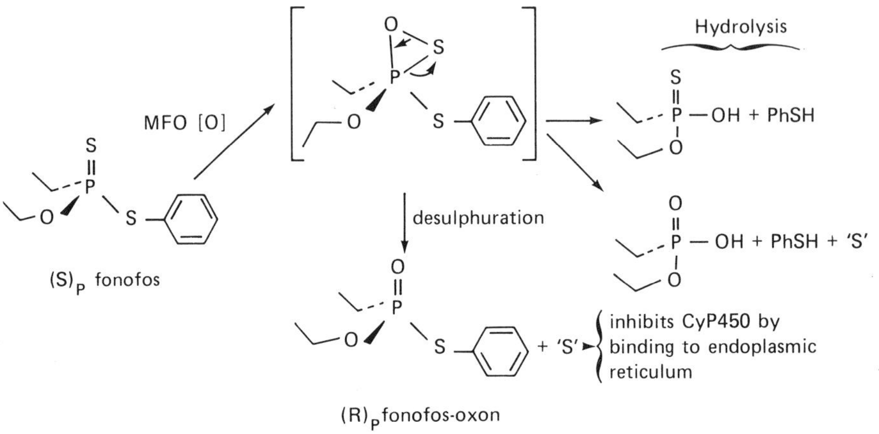

Scheme 1.

Sulphur is concurrently displaced and may inhibit cytochrome P450-dependent oxidations by covalent binding to associated macromolecules. The oxons of the enantiomeric forms of the chiral P=S compounds may be produced and/or metabolized at different rates, with interesting toxicological consequences (Ohkawa, 1982). Thus, the $(R)_p$ enantiomer of the soil insecticide fonofos (**10**) is more toxic (LD_{50}s, 6.3 μg/g and 9.5 mg/kg (oral) respectively) than $(S)_p$fonofos (LD_{50}s, 25 μg/g and 32 mg/kg (oral) respectively) to houseflies and white mice. The $(S)_p$oxon formed from $(R)_p$fonofos with retention of configuration Ohkawa, 1982), is more stable and more toxic to mice (LD_{50}, 6 mg/kg) than the $(R)_p$oxon (LD_{50}, 38 mg/kg) from $(S)_p$fonofos, although the $(R)_p$oxon is the better AChE inhibitor *in vitro* (Lee *et al.*, 1978).

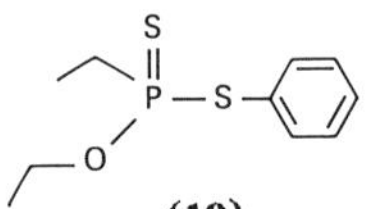

(10)

The difference in toxicity between enantiomers can be large. Thus, the (+) isomer of the contact and stomach poison isofenphos-oxon (**11**), formed from the pro-insecticide isofenphos (P=S), was 83-fold more toxic than the (−) oxon to houseflies (Ueji and Tomizawa, 1986). Further bioactivation of the oxon occurs; the phosphor-oamido-oxon produced by oxidative *N*-dealkylation of isofenphos-oxon was found to be 2300-fold more potent than isofenphos-oxon as an inhibitor of housefly head AChE (Gorder *et al.*, 1986). Chirality may influence the systemic properties of pesticides; (S)$_P$fonofos entered the roots of corn and cotton plants more rapidly than the more toxic (R)$_P$ form but the latter was metabolized more rapidly following absorption (Lee *et al.*, 1980).

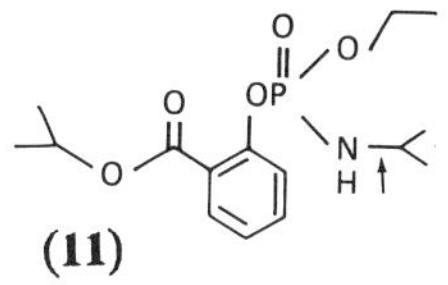

(11)

The replacement of an ethoxy group by an *n*-propylthio group in certain *O*-aryl *O*,*O*-diethyl phosphorothioates gives chiral compounds with favourable mammalian toxicity and improved toxicity to insects, including OP-resistant strains (Hirashima *et al.*, 1984; Drabek and Neumann, 1985). Thus, prothiofos-oxon (**12**) and its *S-n*-butyl homologue are highly insecticidal, compared with the *S*-methyl and *S*-ethyl analogues, especially toward OP-resistant houseflies, although (**12**) and its *S*-methyl, *S*-ethyl and *S-n*-butyl analogues inhibit both susceptible and resistant AChE poorly *in vitro*. In contrast to the *S*-methyl and *S*-ethyl oxons, the *S*-propyl and *S-n*-butyl oxons were converted into more potent inhibitors by rat liver microsomes (+NADPH). This effect appears to result in a change of leaving group for AChE inhibition (Scheme 2); the two higher alkylsulphoxide moieties formed by bioactivation are better leaving groups than 2,4-dichlorophenoxy, whereas the sulphoxides from the smaller alkylthio groups are rapidly detoxified by hydrolysis (Kimura *et al.*, 1984). The (−) enantiomer of the related compound profenofos (**13**; 2Cl,4Br-Ph) is metabolically activated and the less toxic (+) isomer, although a more potent AChE inhibitor *in vitro*, is detoxified by the MFO system (Casida and Ruzo, 1986); a change in leaving group from *O*-aryl to −S(*O*)-*n*-propyl by oxidative bioactivation of (−)-profenofos is believed to result in a phosphorylated enzyme that ages rapidly and irreversibly (Wing *et al.*, 1984) and this may partly explain the efficiency of such compounds against OP-resistant insects.

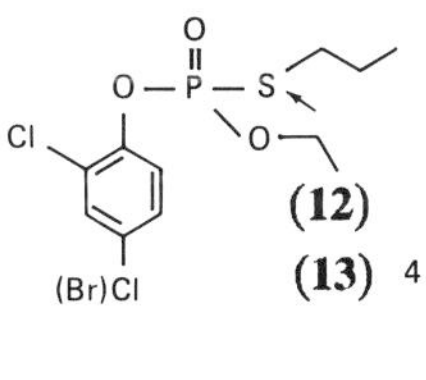

(12)

(13) 4 — Br

The broad spectrum contact and systemic insecticide methamidophos (**14**) is only a moderate AChE inhibitor *in vitro*. However, it has high mammalian toxicity that can be much reduced by conversion into the *N*-acetyl derivative acephate (**15**).

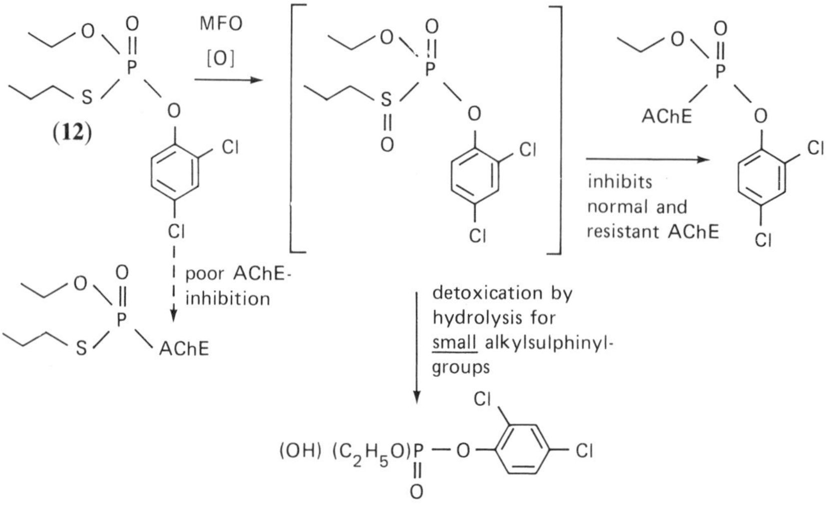

Scheme 2.

Acephate also is a poor AChE inhibitor, which reverts by hydrolysis to (14) *in vivo*, and because of its low toxicity to fish has been extensively used in situations where contamination of water can occur (Green *et al.*, 1984). Methamidophos is thought to be bioactivated by conversion into the corresponding sulphoxide but hitherto this has not been conclusively proven. However, the thiomethyl moiety does appear to be the leaving group for the interaction of methamidophos analogues with both the AChE and neuropathy target esterase (NTE) of hen brain (Vilanova *et al.*, 1987).

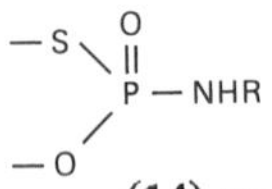

(14) R= H
(15) R= COCH₃

The systemic insectide phosfolan (16) apparently acts as an anticholinesterase *in vivo*, despite the lack of an obvious leaving group and poor intrinsic AChE inhibitory capability. However, it is activated by MFO *in vitro* and can be converted by peracid oxidation into a monosulphoxide and a sulphone (17), which are potent AChE inhibitors. The phosphoramide (18) is generated by microsomal enzymes and, on the basis of chemical hydrolysis results, Casida and Ruzo (1986) suggested that the iminocarbon–sulphur bond in the sulphone (17) may be involved in attack on the active site of AChE. The urine of rats treated with the related compound mephosfolan (19) contained products of oxidation of the methyl group to a carboxyl group. Hydrolytic attack on the P–N bond released the imidodithiolane ring moiety, which underwent ring opening to generate thiocyanate, the predominant tissue metabolite. In cotton plants, mephosfolan was rather stable, a glucoside of the hydroxymethyl derivative being a major metabolite (Menzie, 1980).

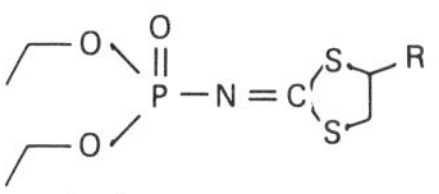

(16) R= H
(19) R= CH₃
(17)
(18) 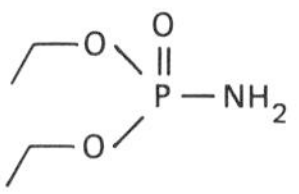

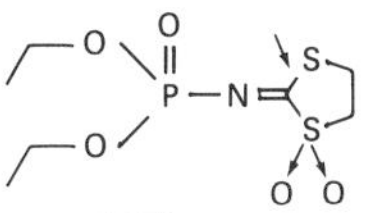

The thiono–thiolo rearrangement of phosphorothionate insecticides can occur during manufacture or storage, resulting in the corresponding phosphate with a thioalkyl instead of an alkoxy group attached to phosphorus. Isomalation (20), the rearrangement product of malathion, inhibits AChE and also the carboxylesterase which detoxifies malathion in mammals (Ryan and Fukuto, 1985). When present at low levels (0.1%) in malathion (9), (20) greatly potentiates this normally very safe insecticide (rat oral LD_{50}, *ca.* 10 g/kg when pure) and has been responsible for episodes of malathion poisoning among spray operators in the tropics. Fenitrothion (21) is activated in the same manner under environmental conditions and the more toxic *S*-methyl oxon produced by this rearrangement also appears in seed coat, meristem and embryo of the white pine treated with fenitrothion (Rosival *et al.*, 1976).

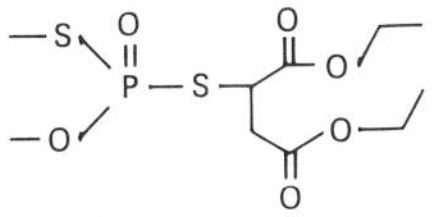

(20)
(21) R= NO₂ 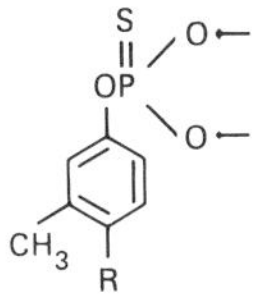
(24) R= S(O)CH₃ (O,O-diethyl)
(22) 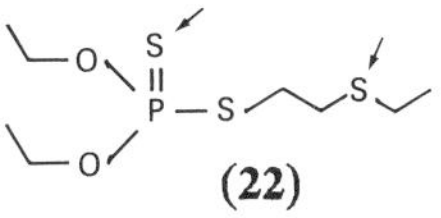
(23) 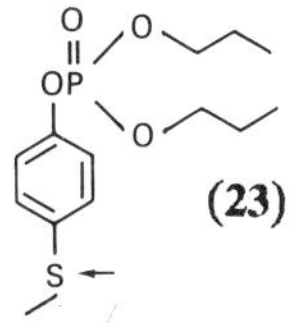

Sulphur-containing structures such as thioether or sulphur heterocycle moieties can be deliberately incorporated into pesticides to give useful properties such as systemic insecticidal activity in plants. The oxidation of such thioalkyl groups is frequently a bioactivation mechanism and is of wide importance in this respect (Brooks, 1984). Thus, the sulphoxide and sulphone from disulfoton (22) are produced as insecticidal systemic metabolites in plants and the corresponding oxons are potent AChE inhibitors (I_{50}s, 10^{-6}–10^{-7} M). For propaphos (23), a systemic insecticide used to control leafhoppers and beetles on rice, conversion of the thiomethyl moiety into the corresponding electron-withdrawing sulphoxide *in vivo* increases the AChE inhibitory activity of the molecule and the sulphoxide produced in rice plants is enriched in the $(R)_S$-(+) enantiomer, an example of the chiral metabolism of an achiral parent molecule (Miyazaki *et al.*, 1985). As an example of differential metabolism that results in selective toxicity between insect species, propaphos was converted into its toxic sulphoxide and sulphone and metabolites arising therefrom in the susceptible green rice leafhopper, *Nephotettix cincticeps*. The common cutworm, *Spodoptera litura*, has an equally sensitive AChE but in contrast to *N. cincticeps* converts propaphos mainly into non-toxic hydrolysis products (Kazano *et al.*, 1983). The aryl sulphoxide derivative fensulfothion (**24**) Worthing and Walker, 1987) is actually used as a persistent soil insecticide and nematicide, as are several other compounds of this general structure. These reactions, like desulphuration, are generally effected by cytochrome P-450-dependent MFO but it may be noted that hepatic FAD-dependent monooxygenase (FMO) also can convert thioether groups into the sulphoxides (Smyser *et al.*, 1985).

3.2.4 Carbamate Insecticides
Like the OP insecticides, the carbamate anticholinesterases offer many possibilities for structural variations in the direction of improved environmental acceptability and have found wide practical use. Methiocarb (**25**) and ethiofencarb (**26**) are sulphur-containing representatives of the widely used phenyl *N*-methylcarbamate structural type. Both are converted into their corresponding toxic sulphoxides and sulphones in plants, animals and soils (see Kuhr and Dorough, 1976, and Brooks, 1984, for references). Aldicarb (**27**), thiofanox (**28**) and methomyl (**29**) are representative of the *O*-(*N*-methylcarbamoyl)oxime class of insecticides (Magee, 1982). Compounds (**27**) and (**28**) are converted into their sulphoxides and sulphones, having greater AChE inhibitory potency than the parents in plants, animals and soil, but methomyl is much more extensively degraded, giving it rather ideal environmental properties in this respect (Kuhr and Dorough, 1976). Aldicarb sulphoxide is fairly stable in plant tissues and appears to be the metabolite mainly responsible for the persistent systemic action, although the slowly formed sulphone is a more potent AChE inhibitor. In fact, this sulphone (aldoxycarb) is used directly as a soil insecticide and nematicide. Approximately 80% of the water-soluble metabolites found in cotton plants at harvest consisted of sugar conjugates of 2-methyl-2-(methylsulphinyl)propanol derived from the sulphoxide (Kuhr and Dorough, 1976).

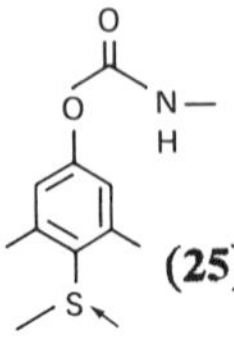

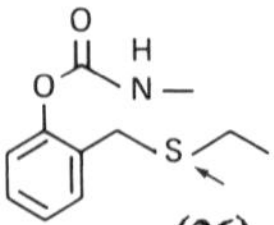

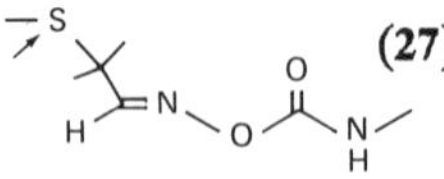

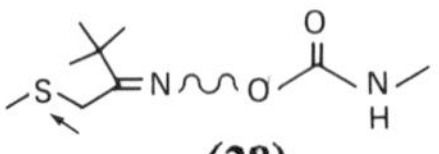

Unfortunately, these valuable insecticides are often highly toxic to vertebrates. However, it has proved possible to improve the selectivity in favour of mammals, while retaining good and sometimes greater insect toxicity than the parent carbamate, by the chemical synthesis of pro-insecticides in which the proton in the CH_3NH- moiety is replaced by a variety of substituents, of which the most useful contain sulphur (Drabek and Neumann, 1985; Umetsu, 1986). The N-substituents investigated have included dimethoxyphosphinothioyl (($(CH_3O)_2P(S))-$) sulphenyl (RS–) and sulphinyl (RS(O)–). In general, the *N*-sulphenyl-*N*-methyl carbamates exhibit lower mammalian toxicity, improved residual insecticidal activity and lower phytotoxicity, compared with the parent carbamates but the systemic activity and stability in storage may be lower. Of these derivatives, carbosulfan (**31**) and furathiocarb (**32**) from carbofuran (**30**), and thiodicarb (**33**) from methomyl (**29**), are now commercial insecticides, while the methomyl derivative U-56295 (**34**) and benfuracarb (**35**) are under development. Compared with methomyl (rat oral LD_{50}, 25 mg/kg), the derivative (**34**) has increased mammalian (rat oral LD_{50}, >8000 mg/kg) and plant safety but retains the toxicity of methomyl towards a range of lepidopteran pests (Dutton *et al.*, 1981).

In general, the selective toxicity to insects seen in these compounds appears to be due to their hydrolysis to the parent toxic carbamates *in vivo* (Marsden *et al.*, 1982; Usui and Umetsu, 1986). In contrast, little of the parent compound is generated in mammals, which mainly detoxify the intact pro-insecticides by other pathways such as sulphoxide and sulphone formation, followed by hydrolytic cleavage of the carbamic ester link (Marsden *et al.*, 1982). The active carbamates and their metabolites are also generated in plant tissues; the excellent systemic activity of benfuracarb (**35**) in cotton is accompanied by N–S bond cleavage to give carbofuran (**30**), which is hydroxylated at C5 and in the *N*-methyl group, followed by the conversion of these metabolites into conjugates (Umetsu *et al.*, 1985).

3.2.5 Other insecticidal compounds

Nereistoxin (**36**), a natural cholinergic insecticide from the marine worm, *Lumbriconereis heteropoda*, contains a disulphide link which has been used to convert nereistoxin into the practical pro-insecticides cartap (**37**), thiocyclam (**38**) and bensultap (**39**). These derivatives are converted into nereistoxin *in vivo* and are generally active against lepidoptera and coleoptera. Cartap is translocated in rice plants and is found in the plant tissues, together with nereistoxin and traces of nereistoxin monosulphoxide (Nishi *et al.*, 1979). Nereistoxin acts at the insect postsynaptic ACh receptor, possibly by covalent binding to it through the corresponding dithiol (Sattelle *et al.*, 1985), from which the above derivatives are prepared. The new nitromethylene heterocycle insecticide (**40**; WL108477), which acts as an ACh agonist at ACh receptors, promises to be well suited for use on rice, combining low mammalian and fish toxicity with high toxicity to rice pests, including planthopper vectors of virus diseases. It is also effective against various insect species that are resistant to OPs and pyrethroids (Harris *et al.*, 1986).

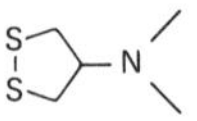

(36) 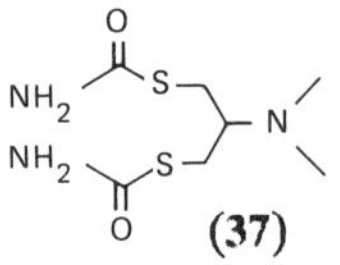   (37) 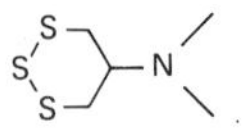   (38)

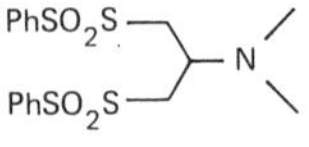

(39) 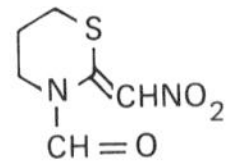   (40) 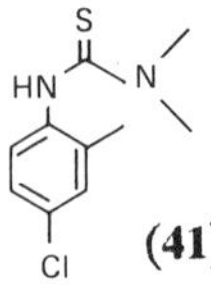   (41)

The acaricide chloromethiuron (**41**) has a similar action to chlordimeform (**42**; CDM), an octopamine agonist in insects which gives effective control of mites and lepidoptera including *Heliothis* sp. on cotton. *N*-sulphenyl-*N'*-arylformamidines analogous to the carbamate derivatives discussed previously have been prepared from the *N*-monodemethyl analogue (**43**; DCDM) of CDM and related compounds, to which derivatives such as the thiobisformamidine (**44**) appear to be converted *in vivo* (Knowles, 1982; Drabek and Neumann, 1985). Structural variants also exist in which the −NCH₃R₂ moiety in (**42**) is incorporated in a thiazolylidene ring (Knowles, 1982).

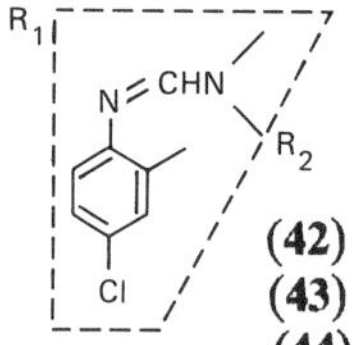

(**42**) CDM, R₂ = CH₃
(**43**) DCDM, R₂ = H
(**44**) R₁ − S − R₁

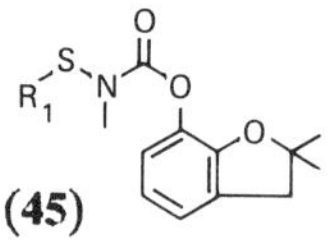

(**45**)

In compounds such as DCDM-*S*-carbofuran (**45**) a formamidine and a carbamate structure are combined in one molecule, which exhibits the activity of each separate moiety. The combination extends the range of pesticidal activity obtained with each parent and an increased effectiveness against both insects and mites, possibly resulting from more favourable toxicokinetic properties of the hybrid molecule. It is likely that in pest species, *in vivo*, the molecule is cleaved on either side of the central sulphur atom to give both the parent carbamate and the formamidine. As in the case of the derivatized carbamates discussed above, improved mammalian safety is believed to result from detoxication initiated by oxidative attack on the intact molecule, in contrast to cleavage resulting in liberation of the parent toxicants in pest insects (Eya and Fukuto, 1986).

The recently introduced larvicide buprofezin (**46**) is probably best regarded as an insect growth regulator, since it inhibits chitin biosynthesis, and also prevents egg deposition by the brown planthopper, *Nilaparvata lugens*, apparently by inhibiting prostaglandin-E₂ biosynthesis (Uchida *et al.*, 1987). Buprofezin is relatively safe to mammals (acute oral LD₅₀ to rats and mice, >2000 mg/kg). It becomes soil bound but disappears with a half-life of 100 days under non-sterile conditions, with 4-hydroxylation of the benzene ring and CO_2 formation as the major route of microbial degradation (Funayama *et al.*, 1986). It is interesting to note that isoprothiolane (**47**), introduced as a fungicide, is an inhibitor of lipid biosynthesis which also has a remarkable insect growth regulating effect on rice planthopper pests. This has been attributed to the inhibition of chitin synthetase in these insects (Uchida *et al.*, 1982).

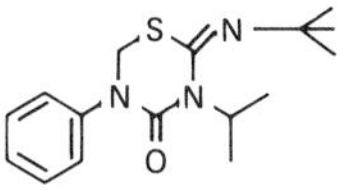

(**46**) 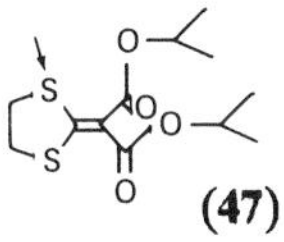   (**47**)

3.3 HERBICIDES

Herbicides, many of them containing sulphur, represent nearly half of the world wide pesticide market, some 200 herbicides and PGR being currently available (Esser, 1986).

3.3.1 Thiocarbamate herbicides

Thiocarbamates are widely used as pre- and post-emergence herbicides. Compounds of this class include thiobencarb (**48**), orbencarb (**49**), molinate (**50**), EPTC (**51**) and diallate (**52**), which are metabolically converted into sulphoxides that are generally more phytotoxic, though less biologically stable, than the parent compounds. They tend to be selectively active on weeds, although the sulphoxide from tiocarbazil (**53**) is toxic to both rice and barnyard grass (Santi and Gozzo, 1976).

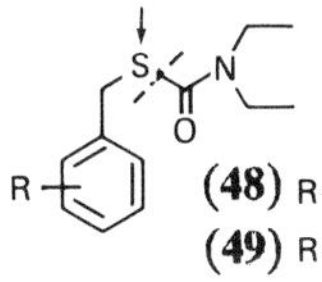

(**48**) R= 4 —Cl   (**50**)
(**49**) R= 2 —Cl

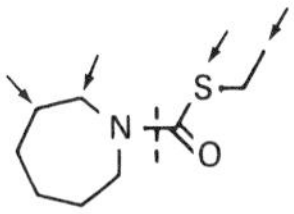

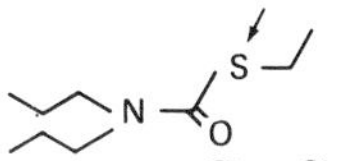

(**51**)

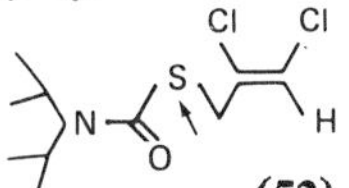

(**52**)

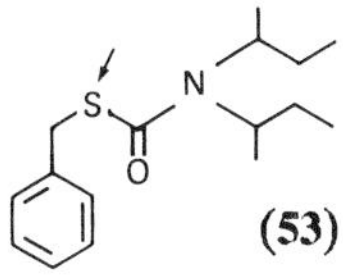

(**53**)

Thiobencarb (**48**) (benthiocarb), which inhibits the biosynthesis of long chain fatty acids in both plants and insects (Brown, 1987), is used as a herbicide in rice paddies, while its 2-chloro isomer, orbencarb (**49**), is highly effective against weeds in crops such as wheat and soybean. Both compounds gave sulphoxides in soybean and followed quantitatively similar pathways thereafter; for orbencarb sulphoxide cleavage gave either 2-chlorobenzoic acid, via 2-chlorobenzyl alcohol, each present both free and conjugated, or 2-chlorobenzylsulphonic acid. Alternatively, hydrolytic cleavage without sulphoxide formation gave 2-chlorobenzylthiol, which was methylated, and the thiomethyl derivative subsequently oxidized to the corresponding sulphoxide and sulphone.

N-dealkylation and hydroxylation of the 2-chlorobenzene ring in the 4 and 5 positions were also observed. The same metabolites were formed in soil, with major hydrolytic cleavage of the $-S-C=O$ bond to give 2-chlorobenzylsulphonic acid and methyl 2-chlorobenzylsulphone as end-products (Ikeda *et al.*, 1986a, b). The herbicidal activity, if any, of these sulphur-containing metabolites is not indicated. Microbial populations from soil can dechlorinate thiobencarb under anaerobic conditions to give 4-dechlorothiobencarb, which causes dwarfing of rice plants (Moon and Kuwatsuka, 1985, 1987).

Molinate (**50**) is a selective herbicide used to control barnyard grass (*Echinochloa* sp.) in rice cultures; its metabolic pathways are summarized in Scheme 3. More

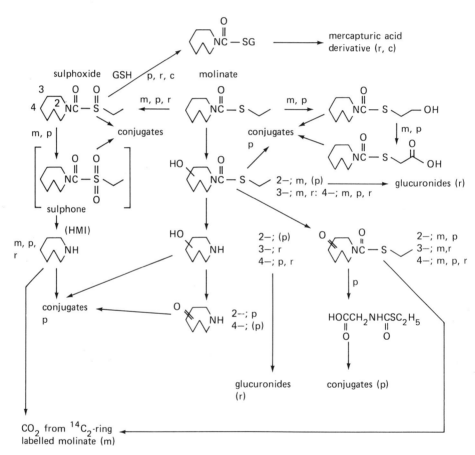

m = microorganisms; p = rice, barnyard grass; r = rat, c = Japanese carp
( ) formation of metabolite inferred from occurrence of earlier or
later products

Scheme 3.

molinate was absorbed by barnyard grass and converted into metabolites than was the case with rice (Imai and Kuwatsuka, 1984). The metabolites identified in both plants, with their conjugation products in some cases, were molinate sulphoxide, the azepine ring-4-hydroxyl, -2-oxo, and -4-oxo derivatives, the S-2-hydroxyethyl and S-carboxymethyl derivative, hexamethyleneimine (HMI), 4-OH-HMI, 2-oxo-HMI and S-ethyl N-carboxymethylthiocarbamate. Apart from the formation of a greater proportion of conjugates derived from basic (imino) compounds in barnyard grass than in rice, which produced more glucosides of acidic and neutral metabolites, there was no obvious mechanism of selectivity. Similar degradation pathways, except for conjugation, were observed in non-sterile soil, with sulphoxide formation followed by hydrolysis to HMI, azepine ring oxidation under upland conditions (aerobic) and oxidation of the S-ethyl moiety under flooded (anaerobic) conditions. Degradation to CO_2 was more rapid under upland conditions (Thomas and Holt, 1980; Imai and Kuwatsuka, 1982) and several strains of microorganisms have been isolated which degrade molinate by co-metabolism in the presence of a nutrient source (Imai and Kuwatsuka, 1986). In rats, excretion of the mercapturic acid derivative as the major metabolite in urine indicated conversion to the sulphoxide, followed by conjugation with GSH, to be the major biotransformation pathway (De Baun *et al.*, 1978) and this route is also prominent in the Japanese carp (Lay and Menn, 1987).

EPTC (**51**) and diallate (**52**) are believed to inhibit gibberellic acid precursor biosynthesis (Wilkinson, 1986) and the sulphoxides, which readily transfer their carbamoyl groups to GSH and other thiols, may play an active role in this action. This view is supported by the observation that the herbicide antidote N,N-diallyl-2,2-dichloroacetamide (R-25788) protects corn from EPTC injury by increasing both the synthesis of GSH and the level of GSH S-transferase (GST) activity, thereby stimulating the detoxication of EPTC sulphoxide through conjugation of its carbamoyl moiety (Casida and Ruzo, 1986). However, according to Horvath and Pulay (1980), the corresponding sulphone is much more reactive than the sulphoxide toward thiol groups. The above may not be the only mechanism for the observed protective effect (Esser, 1986), which is achieved in corn with only 8% of the antidote incorporated into EPTC formulations, although the protective efficiency varies between different corn lines (Lay and Niland, 1985). Molinate and its sulphoxide and sulphone are toxic to the Japanese carp and this presents problems in rice cultivation. The protective effect seen with R-25788 and the related compound N-ethyl-N-benzyldichloroacetamide is associated with elevated fish liver GSH levels and enhanced excretion of molinate mercapturic acid in the bile, without an apparent increase in GST activity (Lay and Menn, 1987).

Diallate (**52**) is converted into potent mutagens by mammalian MFO, and the formation of the mutagenic aldehyde 2-chloroacrolein during this process supports the view that a transient sulphoxide is formed which cleaves, with rearrangement, to give diisopropylcarbamoylsulphenyl chloride and the mutagenic aldehyde (Marsden and Casida, 1982) (see Chapter 4, this volume, Part B).

3.3.2 Triazines and Triazones

Ashton and Glenn (1979) noted that prometryne (**54**) was a more effective inhibitor of photosynthesis and of RNA, lipid and protein synthesis than the corresponding triazines having Cl or CH_3O- substituents. This may relate to the formation of a sulphoxide from the thiomethyl group which is a unique feature of prometryne and cyanatryn (**55**).

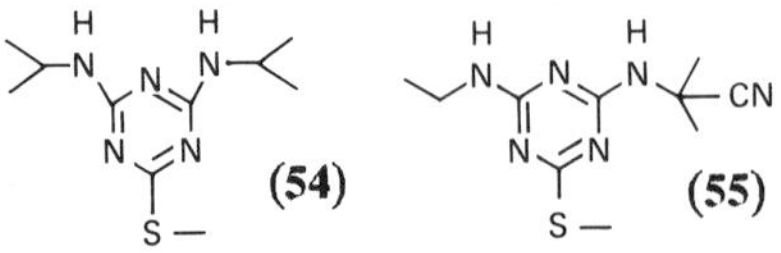

(**54**)   (**55**)

Metribuzin (**56**) is a widely used pre- or post-emergence triazinone herbicide, which inhibits photosynthesis, apparently by binding to the same receptor site in the electron transport pathway of photosystem II as the s-triazines (Draber and Fedtke, 1979). The metabolites detected suggest that an enzymic sulphoxidation in plants (Frear *et al.*, 1985) and sulphoxidation both with and without deamination in mammals (Climie and Hutson, 1979; Bleeke *et al.*, 1985) introduce a reactive $-S(O)CH_3$ moiety which can be displaced by GSH or protein thiol groups, as in the case of the thiocarbamate sulphoxides or cyanatryn. Metribuzin has a low acute oral toxicity to mammals (acute oral LD_{50} to rats, 2200 mg/kg) but an acute lethal dose given to mice caused hepatotoxicity, associated with depletion of GSH and binding of the triazinonyl group to blood and liver proteins. The protection against this effect afforded by pretreatment with piperonyl butoxide (PBO) and the identification of mercapturate conjugates linked through the original site of attachment of the thiomethyl group supports the view that sulphoxidation occurs, although the sulphoxide and sulphone have so far only been identified as products of peracid oxidation of metribuzin (Casida and Ruzo, 1986).

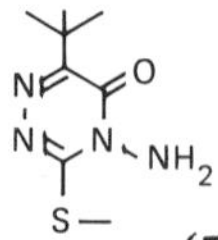

(**56**)

3.3.3 Sulphur heterocyclics, sulphones, sulphonates, sulphonamides

Some urea-type herbicides contain a sulphur heterocycle. Examples are benzthiazuron (**57**) and methabenzthiazuron (**58**), which are selective, and ethidimuron (**59**), used for total weed destruction in non-cropped areas. Ethidimuron and nitralin (**60**) contain an alkyl sulphonyl group, which frequently apperas in herbicides as a substituent on carbon, as does the sulphonate moiety, either substituted, as in ethofumasate (**61**), or free, as in the sulphonic acid derivative 2,4-DES (**62**). The latter compound is a proherbicide, which is hydrolysed in moist soil to the corresponding ethanol derivative, a precursor of 2,4-D, to which it is converted by oxidation. The bis-sulphone dimethipin (**63**) is used as a defoliant for cotton, nursery stock, rubber trees and vines (Worthing and Walker, 1987).

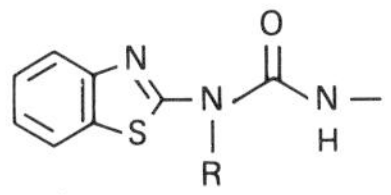

(57) R= H
(58) R= CH$_3$

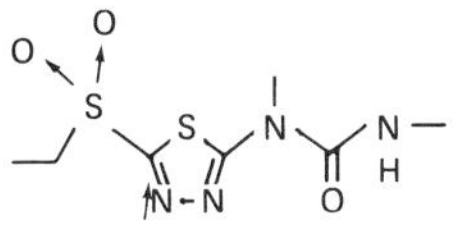

(59)

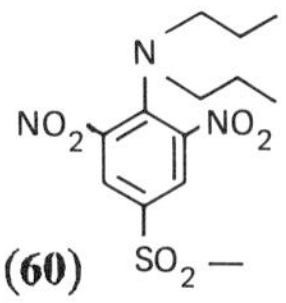

(60)

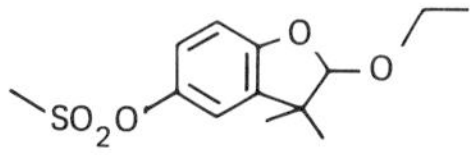

(61)

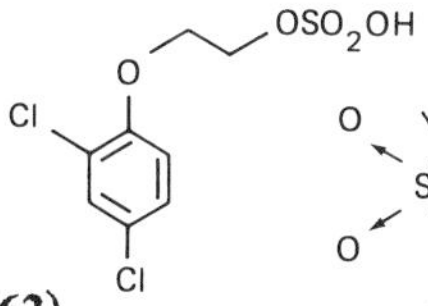

(62)

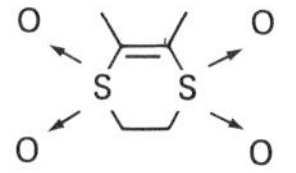

(63)

The contact herbicide bentazon (64), which acts on photosystem II, also contains a fully oxidized sulphur atom. Its selective toxicity appears to depend mainly on differential detoxication; rapid hydroxylation at C6 followed by conjugation with glucose at this carbon occur in rice, whereas (64) is very stable in the weed purple nutsedge (Mine *et al.*, 1975). Soil microorganisms also effect the C6-hydroxylation of bentazon under aerobic conditions but the product is rapidly incorporated into humic substances as an insoluble bound residue and cannot be isolated (Otto *et al.*, 1979). Other selective herbicides containing fully oxidized sulphur are perfluidone (65), which is believed to act as an uncoupler of photosynthetic phosphorylation by perturbing the thylakoid membrane, and the exceptionally active sulphonylurea derivative chlorsulfuron (66), a plant growth inhibitor used to control broadleaf weeds in cereals and believed to inhibit branched chain amino acid biosynthesis at the acetolactate synthase level (Schloss *et al.*, 1988). Tolerance in grasses and broadleaf plants has been linked with hydroxylation of the aromatic ring and of the *s*-triazine methyl substituent respectively of chlorsulfuron, followed by carbohydrate conjugation of these metabolites (Hutchison *et al.*, 1984).

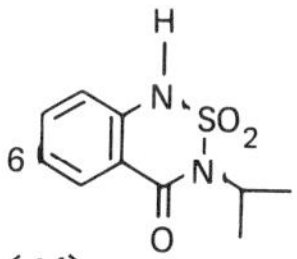

(64)

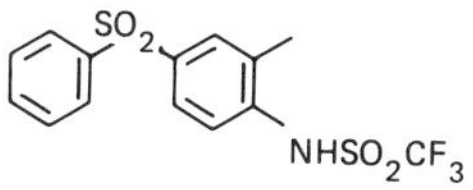

(65)

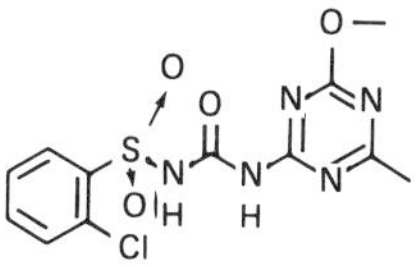

(66)

The diphenylether herbicide fomesafen (67) is a member of the group believed to induce chloroplast membrane damage by lipid peroxidation, following activation by light (Orr and Hess, 1981). The presence of the methanesulphonylcarboxamide moiety confers greater selectivity in favour of soybean and greater residual control of broadleaf weeds, compared with the corresponding molecules' with a free carboxyl group (Colby *et al.*, 1983). As in the cae of the anilide-type herbicide mefluidide (68), the sulphur-containing moiety appears to be acting as a useful derivatizing group, rather than by changing the basic mechanism of action.

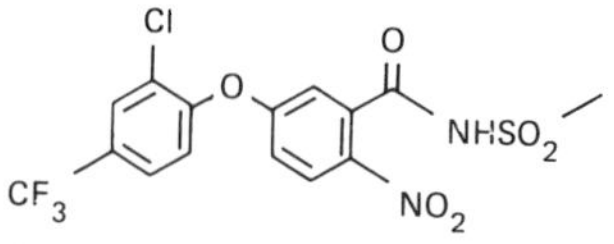

(67) 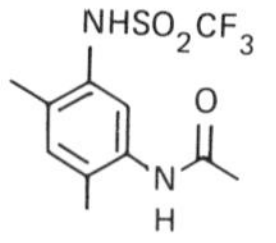   (68)

Sulphonamides inhibit plant growth by interrupting the biosynthesis of folic acid, as they do in microorganisms. The chlorosis-inducing carbamate herbicide asulam (**69**) is structurally similar to sulphanilamide and the herbicidal effects of both were reversed by 4-aminobenzoic acid, indicating a similar action through the inhibition of folic acid biosynthesis (Veerasekaran *et al.*, 1981). Another sulphonamide, oryzalin (**70**), disrupts plant cell division by interfering with the synthesis of microtubule subunits (Draber and Fedtke, 1979).

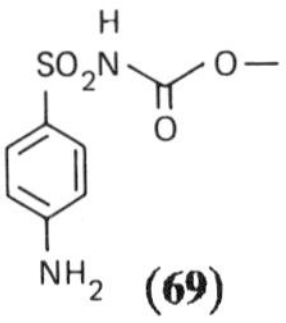

(69) 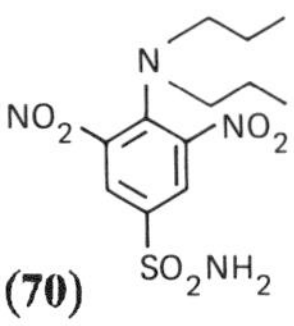   (70)

3.3.4   Herbicidal organophosphorus esters

A number of herbicidal phosphorothionate esters are known which apear to be derivatives of known herbicidal types such as dinitrophenols, substituted *s*-triazines or urea type herbicides, to which they should revert by hydrolysis in plant tissues (Brooks, 1984). Others have activity in their own right. Examples are *S,S,S*-tributylphosphorotrithioate, a well-established defoliant for cotton and the phosphorothionate piperophos (**71**), which is used in combination with the thiomethyl s-triazine herbicide dimethametryn (**72**) Worthing and Walker, 1987) to control grasses and sedges in flooded rice fields. In rice plants and in rats (Scheme 4), piperophos underwent hydrolytic P–S bond cleavage, both before and after desulphuration to the phosphate. The liberated thioacetyl piperidine moiety was then oxidized intact to the sulphonic acid and also methylated, then oxidized to the corresponding sulphoxide in rice (Mayer *et al.*, 1981), whereas a variety of piperidine ring hydroxylated or piperidine ring cleaved sulphoxides and sulphones were produced from this fragment in the rat (Esser, 1986). Another OP herbicide, amiprophos-methyl (**73**) is a cell division inhibitor which appears to prevent tubulin formation (Draber and Fedtke, 1979).

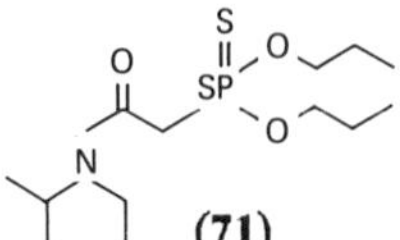

(71) 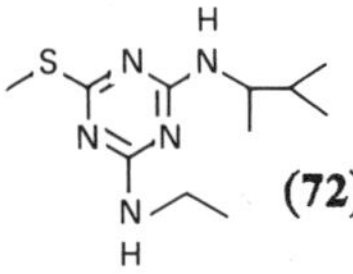   (72) 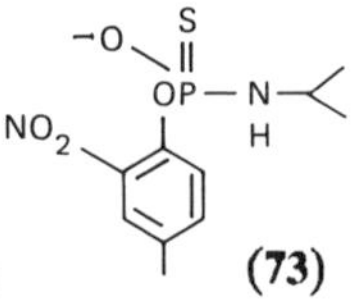   (73)

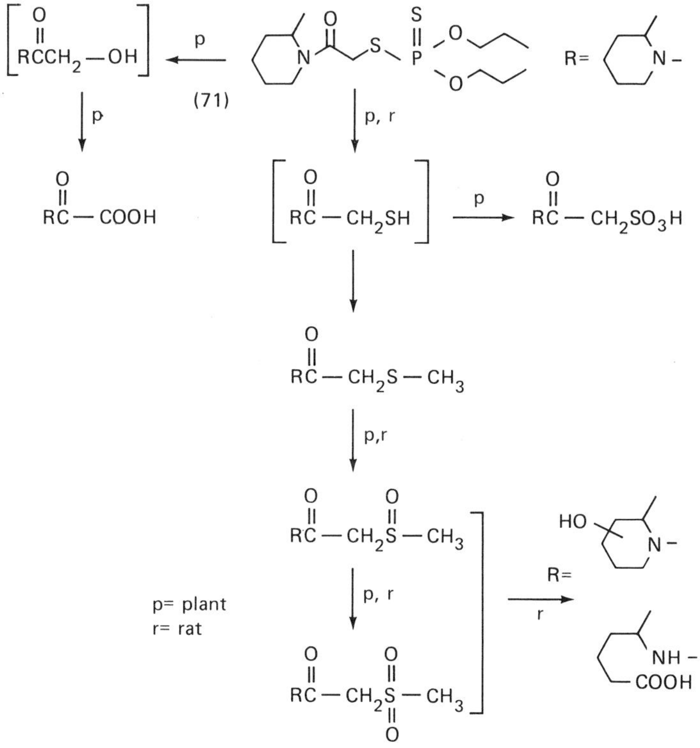

$$\left[\begin{array}{c} O \\ \parallel \\ RCCH_2-OH \end{array} \right] \xleftarrow{\quad p \quad}$$ (71)

R= (piperidine structure with N–)

$$\downarrow p$$

$$\begin{array}{c} O \\ \parallel \\ RC-COOH \end{array}$$

$$\left[\begin{array}{c} O \\ \parallel \\ RC-CH_2SH \end{array} \right] \xrightarrow{\quad p \quad} \begin{array}{c} O \\ \parallel \\ RC-CH_2SO_3H \end{array}$$

$$\downarrow$$

$$\begin{array}{c} O \\ \parallel \\ RC-CH_2S-CH_3 \end{array}$$

$$\downarrow p,r$$

$$\left[\begin{array}{cc} O & O \\ \parallel & \parallel \\ RC-CH_2S-CH_3 \end{array} \right]$$

HO (piperidine structure with N–)

R=

$$\downarrow p,r$$

r

(piperidine structure with NH– COOH)

p= plant
r= rat

$$\begin{array}{cc} O & O \\ \parallel & \parallel \\ RC-CH_2S-CH_3 \\ & \parallel \\ & O \end{array}$$

Scheme 4.

Oxidative desulphuration is prominent in the reactions of some of the phosphorothionates and may aid subsequent hydrolysis of the esters. It is also evident in the rapid conversion of the pro-herbicide chlorthiamid (**74**) into the uncoupling agent dichlobenil (**75**) in soil, which presumably occurs via oxidation to the corresponding amide, followed by dehydration.

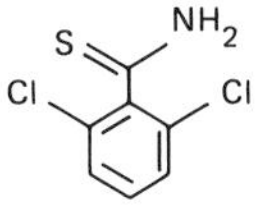

(**74**) 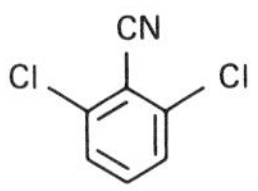   (**75**)

3.4 FUNGICIDES

Fungicides follow insecticides in order of worldwide pesticide sales. They constituted about 20% of the total market value of pesticides in 1983 (Somerville, 1986) with a value of $2.8 billion. Sulphur-containing compounds are prominent, the non-systemic dithiocarbamates such as mancozeb and maneb being most extensively used in agriculture. These compounds constituted about 37% of the sales of non-systemic fungicides, which were 58% of total fungicide sales. Next came phthalimides such as captan (16% of non-systemic fungicides; 9% of total sales). Although the metabolic pathways of fungicides in plants have been extensively studied, there is a lack of detailed information about the enzymes involved. For reviews see Langcake *et al.* (1983), Vonk (1983) and Somerville (1986).

3.4.1 Carbamates containing sulphur

Thiram (**76**) is representative of the *N*-dimethyldithiocarbamate group. In plants, (**76**) is cleaved at the disulphide link and the *N*-dimethyldithiocarbamoyl ion is then conjugated with glucose and incorporated into amino acids and proteins (Vonk, 1983). Thiram and other dithiocarbamates are metabolic poisons with acute effects similar to those of carbon disulphide, which may arise, with dimethylamine, from first-formed *N,N*-dimethyldithiocarbamic acid. However, in contrast to CS_2, thiram causes thyroid dysfunctions in vertebrates, which is presumed to result from the release of elemental sulphur in the folicular cells (World Health Organisation, 1985).

(**76**)

The *N*-dimethyldithiocarbamate group is much less important than the ethylene bisdithiocarbamate group, represented by the related compounds zineb (**77**), maneb (**78**) and Mancozeb (**79**). These are insoluble, polymeric coordination complexes of the bisdithiocarbamate moiety with Zn, Mn and Zn–Mn respectively which degrade readily in water to fungitoxic products that also appear in treated plants. Ethylene diamine and ethylenethiourea (**80**; ETU) are formed in both animals and plants, and plants produce ethylene bisisothiocyanate and EBS (**81**), together with ETU and other compounds (**82–85**) derived from it (Vonk, 1983). The picture is complicated by the formation of elemental sulphur and hydrogen sulphide. EBS is more fungitoxic than the parent sodium salt (nabam) and ethylenethiouram mono- and disulphides (**86**) are also possible fungitoxic conversion products in plants. ETU (rat, oral LD_{50}, 900 mg/kg), which is of toxicological concern as a known mammalian carcinogen and teratogen, is believed to arise as a contaminant of the fungicidal formulations rather than as a plant metabolite. Ethylene thiuram monosulphide (**86**) appeared together with ETU in the urine and faeces of rats dosed with ^{14}C-maneb, and ETU in the urine of maneb-treated mice (Somerville, 1986). ETU is relatively stable to hydrolysis but very susceptible to oxidation to ethylene urea (**82**). It is microbially oxidized to CO_2 in soil and Engst (1979) concluded that there is no reason for concern about possible contamination of the environment by ETU when access to oxygen is adequate.

[—SCS.NHCH$_2$CH$_2$NHCS.S—M—]$_x$ [—SCS.NHCH$_2$CH$_2$NHCSS.Mn]$_x$

(77) M= Zn **(78)** M= Mn   **(79)**   (Zn)$_y$

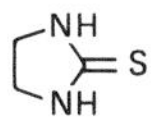

(80) 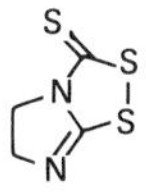 **(81)** 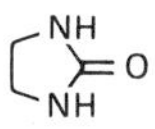 **(82)** 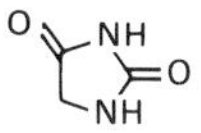 **(83)** 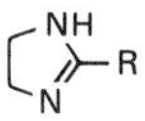 **(84)** R= H   R=—N

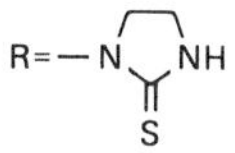

(85)

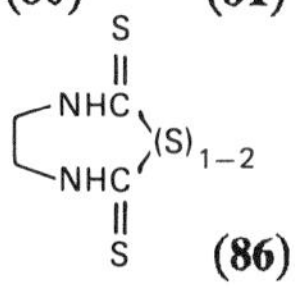

(86)

The soil applied systemic fungicide prothiocarb **(87)** is believed to be active against some fungi through the generation of ethanethiol and hence L-ethionine, an antimetabolite of L-methionine, which can reverse the action in such cases (Kerkenaar and Kars Sijpesteijn, 1977). In other cases, an effect of the intact molecule on the permeability of fungal membranes is indicated (Langcake *et al.*, 1983).

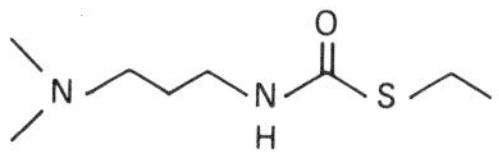

(87)

3.4.2 Sulphenamides

These extensively used fungicides include the phthalimides, captan **(88)** and folpet **(89)** containing the N–SCl$_3$ moiety, captafol **(90)**, and compounds containing the N–SCCl$_2$F moiety, dichlofluanid **(91)** and tolylfluanid **(92)**. There is little information about the metabolism in plants of these widely used compounds, which have been regarded as carriers of the corresponding haloalkylthiols that can be generated from them by N–S bond cleavage. The liberated thiols can then be converted into H$_2$S, HCl and COS and toxic thiophosgene (CSCl$_2$) and its derivatives by interaction with cellular thiols. This is substantiated by the appearance in fungal spores of thiazolidine-2-thione-4-carboxylic acid **(93)**, a condensation product of cysteine with CSCl$_2$, and a similar derivative of glutathione, following treatment with captan. While it is by no means certain that these reactive intermediates are solely responsible for fungitoxicity (Schuphan and Casida, 1983), their destruction of GSH and covalent binding to cellular constitutents at a multiplicity of fungal target sites (Siegel, 1970) constitute a 'shot gun' attack which should aid fungicidal efficiency and hinder the development of resistance. The formation of thiophosgene as a metabolite of captan in rats was indicated by the appearance of the thiazolidine acid **(93)** in urine, together with dithiobis(methanesulphonic acid) **(94)** and its monosulphoxide. The last two compounds became radiolabelled when rats were dosed simultaneously with captan and ^{35}S-sulphite, demonstrating that thiophosgene is trapped by reacting with endogenous sulphite (De Baun *et al.*, 1974).

(88) R= $SCCl_3$
(90) R= $SCCl_2CHCl_2$

(89)

(91) R= H
(92) R= CH_3

(93)

(94)

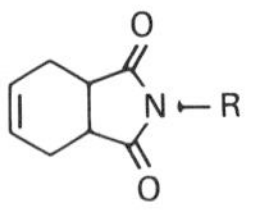

3.4.3 Benzimidazoles

Benzimidazoles are the largest group of systemic fungicides in terms of market size but they also have veterinary uses as anthelmintics and medical uses as antitumour agents (see Volume 3 of this series). The discovery of the fungicidal activity of thiabendazole (**95**) three years after its introduction as an anthelmintic led to the rapid development of benzimidazoles for use against a range of economically important crop diseases. The carbamate derivatives thiophanate (**96**) and thiophanate methyl (**97**) are fungitoxic by virtue of their conversion into the potent benzimidazole derivatives EBC (**98**) and MBC (**99**; carbendazim) respectively in plants. This conversion explains the similarity between the actions of these compounds and that of thiabendazole, which involves the inhibition of fungal mitosis by preventing the normal assembly of microtubule subunits. As precursors of EBC and MBC the thiophanates are therefore regarded as profungicides. In plants, carbendazim was successively converted into 2-aminobenzimidazole and benzimidazole, the latter being further converted into aniline and 2-aminobenzonitrile; thiabendazole gave benzimidazole 2-carboxyamide by thiazole ring cleavage and also benzimidazole but some of these transformations may be abiotic. Similarly, the formation of MBC from thiophanate-methyl may be partly enzymic and partly photochemical (Langcake *et al.*, 1983; Vonk, 1983).

(95)

(96) R= C_2H_5
(97) R= CH_3

(98) R= C_2H_5 (EBC)
(99) R= CH_3 (MBC)

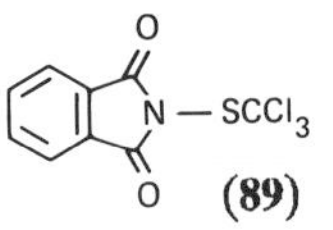

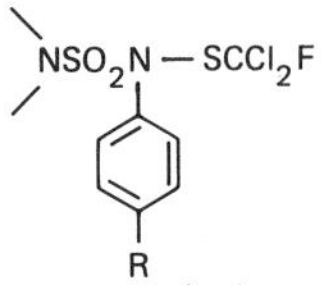

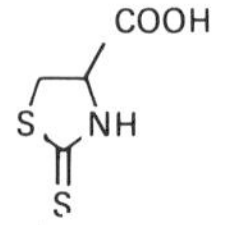

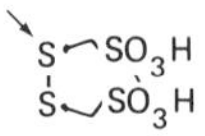

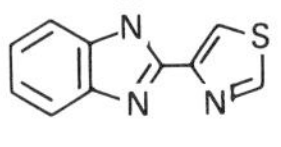

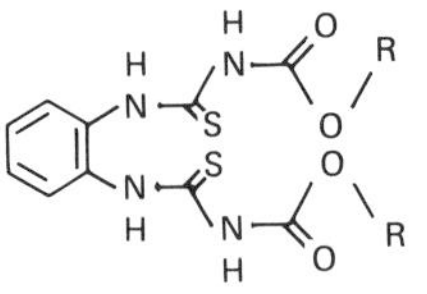

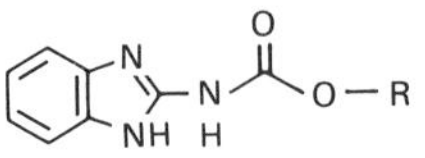

3.4.4 Phosphorothiolates and other fungicides

The activities of a number of phosphorus esters containing a latent fungitoxic component such as thiophenol, 4-methylthiophenol or 4-methoxythiophenol appeared to correlate simply with those of the thiophenols expected to be liberated by hydrolysis of the esters (Eto *et al.*, 1975). Phosphorus esters which have found practical application are the phosphorothiolate (PTL) fungicides edifenphos (**100**), iprobenfos (IBP; **101**) and EBP (**102**), and pyrazophos (**103**). The PTL fungicides are effective against rice blast disease caused by the fungus *Pyricularia oryzae* and appear to affect cell membrane integrity by inhibiting the conversion of phosphatidyl ethanolamine into phsophatidyl choline, mediated by adenosylmethionine methyl transferase (Yamaguchi, 1982; Kurogochi *et al.*, 1985).

Comparing the metabolism of edifenphos in three strains of *P. oryzae* having varying sensitivities to the compounds, Kurogochi *et al.* (1985) concluded that P–S bond cleavage, effected by fungal MFO, occurred extensively only in the most susceptible strain and is a bioactivation mechanism, since the P–S bond cleavage product, benzenethiol, although only poorly active *in vivo*, was a more potent inhibitor of phospholipid *N*-methyltransferase than edifenphos *in vitro*. However, the metabolite formed by mono-4-hydroxylation of (**100**) was fungitoxic, even to a resistant strain, indicating another possible bioactivation mechanism. A critical factor may be the ability of the parent OP fungicides to liberate these metabolites at appropriate sites within the cell membrane, a property which is difficult to test *in vitro*. The mode of action of pyrazophos (**103**), a systemic fungicide used to control powdery mildews on a variety of crops, is unclear. Hydrolysis of the phosphorus ester group, which may be accelerated by desulphuration to give the more liabile oxon, is thought to be a bioactivation reaction, leading to the fungitoxic 2-hydroxy derivative (see structure **103**), which is not formed in certain insensitive fungi, including resistant strains of *P. oryzae*.

It is interesting to note that, although structurally distinct from the PTL fungicides, isoprothiolane (**47**), another fungicide used to protect rice, appears to have a similar effect on lipid biosynthesis. The incorporation of acetate into fatty acids and triglycerides and the incorporation of L-[methyl-^{14}C]methionine into phospholipids were markedly decreased (Kakiki and Misato, 1979). The monosulphoxide is readily formed in rats and very slowly in rice plants. Interestingly, rat liver 9000*g* supernatant gave the racemic sulphoxide, whereas the microsomal fraction gave an enrichment of the (+) enantiomer. Both enantiomers were racemized by rat liver cytosol and in rice plants, with an attendant reduction to isoprothiolane (Uchida *et al.*, 1986). Isoprothiolane has a useful additional attribute as an insect growth regulator for major planthopper pests of rice (Uchida *et al.*, 1982).

Several fungicides contain a sulphonate, sulphamate or sulphamyl group, either overtly or in derivatised form. The soil fungicide fenaminosulf (**104**) interferes with the complex I region of the fungal respiratory chain, at the level of the flavin component of NADH dehydrogenase. Bupirimate (**105**), the dimethylsulphamate ester of ethirimol (**106**), is readily converted into ethirimol, a fungicide specifically active against certain powdery mildews on foliage and fruit. Further metabolism then follows the ethirimol route, which in plants involves conjugation of the hydroxyl group with glucose, *N*-dealkylation, and hydroxylation of the *n*-butyl substituent. Ethirimol is believed to inhibit the enzyme adenosine deaminase, which converts adenosine into inosine during the early stages of mildew development (Holloman and Chamberlain, 1981). Probenazole (**107**) is an indirectly acting systemic protec- tant for rice against the blast disease caused by *P. oryzae*; it decomposes into allyl 2- sulphamylbenzoate in water and in rice plants is converted into saccharin and its *N*- glucosyl conjugate. These are not fungitoxic but may be active in stimulating the plant's natural defences against fungal invasion (Yamaguchi, 1982).

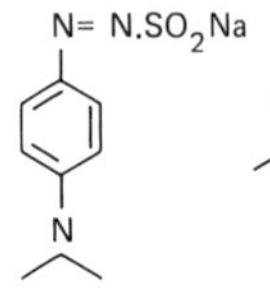

(**104**) 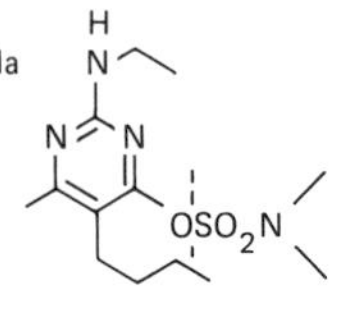   (**105**)   (**107**)
                    (**106**) OH

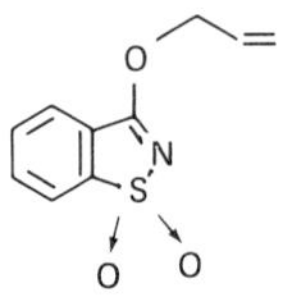

Rice blast disease is also prevented by tricyclazole (**108**), a potent systemic protectant which supresses melanin formation by *P. oryzae* at the level of 1,3,8- trihydroxynaphthalene conversion into 3,8-dihydroxytetralone (vermalone). The production of the 1,8-dihydroxynaphthalene required for melanin synthesis is thereby prevented (Woloshuk and Sisler, 1982).

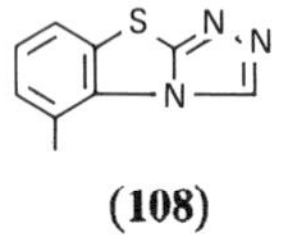

(**108**)

Carboxin (**109**), used as a seed treatment for cereals against bunts and smuts, and its sulphone (**110**) inhibit respiration at the succinate–ubiquinone level in the electron transport systems of fungi, bacteria and mammals. The sulphur atom is not essential, since it is not present in later members of the group, but its oxidation to the sulphoxide and small amounts of the sulphone in plant tissues may aid fungicidal efficiency. The sulphone, also used as a fungicide, is a more stable fungitoxicant than the parent chemical in plants and soil and is translocated in a similar way. A soil microorganisn, *Pseudomonas aeruginosa*, oxidized carboxin to the sulphoxide and then to the sulphone which hydrolysed with ring opening and deacetylation to give 2-

(vinylsulphonyl)acetanilide, an intermediate in the ultimate formation of sulphate, aminophenols and ammonia (Balasubramanya *et al.*, 1980).

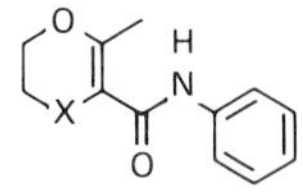

(**109**) X= S
(**110**) X= SO$_2$

Buthiobate (**111**) is a highly effective fungicide for the control of powdery mildews and a member of the remarkable group of ergosterol biosynthesis inhibitors which act by blocking the cytochrome P-450 involved in the C14-demethylation of lanosterol. The $-S(O)$-*n*-butyl derivative formed by monosulphoxidation of (**111**) has been detected in plants together with the pyridyl thiocarbamate (**112**), bis(4-*t*-butylbenzyl)disulphide and 4-*t*-butylbenzaldehyde (Ohkawa *et al.*, 1976a). Sulphoxidation is referred to as a bioactivation (Miyamoto and Ohkawa, 1979), which may be true in the sense that the sulphoxide is also fungitoxic. However, the heterocyclic ring is known to be critically involved in the interaction of these inhibitors with cytochrome P-450 and the parent compound is presumably fungitoxic *per se*. Major urinary metabolites in the rat were derived from hydrolysis of the primary metabolite formed by hydroxylation of one methyl group in the *t*-butyl moiety. The benzylthiol derivative released thereby was methylated and then underwent further oxidation to the methylsulphinyl carboxylic acid (**113**). Excretion of the pyridino fragment as a cyclized conjugate of cysteine (**114**) implies that the hydroxylated primary metabolite, or the sulphoxides formed from it, are attacked initially by GSH (Ohkawa, *et al.*, 1976b).

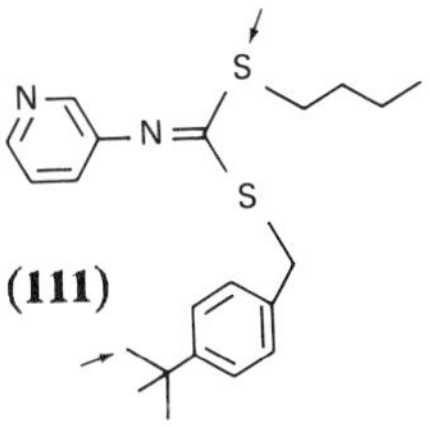

(**111**) 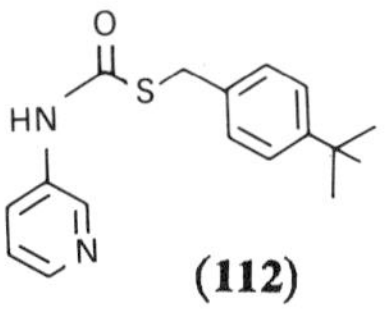   (**112**)

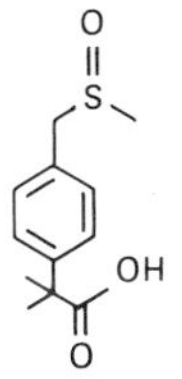

(**113**) 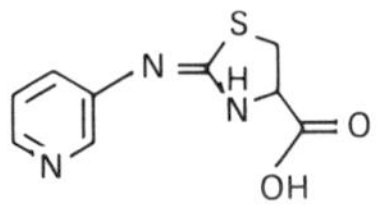   (**114**)

3.5   METABOLISM AND INTERACTIONS BETWEEN CHEMICALS

Although the metabolic pathways of xenobiotics, especially pesticides, have been extensively studied in plants and microorganisms, rather little is known about the enzymes involved in their metabolism. Some plant hydroxylases have the properties of MFO but these may be involved to only a limited extent in xenobiotic biotransformation; peroxidases occur widely in plant tissues and have been implicated in metabolism (Dennis and Kennedy, 1986). This dearth of information applies to the enzymes involved in desulphuration and in the formation of sulphoxides and sulphones, which are the reactions of major interest in this chapter. The enzymes of microbes and plants appear to differ from those which catalyse the corresponding reactions in animals and more work is required to clarify their biochemistry.

In the context of enzymology, much interest centres on the synergistic effects that are apparent between OP compounds and other pesticides and these may have both disadvantages and practical uses. Some OPs potentiate anilide herbicides in rice by inhibiting the rice aryl acylamidase which detoxifies them. However, the inhibition of this enzyme in crabgrass gives a useful synergistic effect which has been used to enhance control of the weed. For OP insecticides, the effect appears to reside in the phosphate analogues, so that P=S compounds must be desulphurated *in vivo*, as in the case of AChE inhibition. Phosphorothiolate (PTL) fungicides such as iprobenfos (**101**) also inhibit this enzyme and their fungicidal action may in turn be synergized by anilide herbicides.

Synergistic interactions between chemicals on insects have long been used to explore the nature of enzymic biotransformations. Thus, iprobenfos, previously mentioned, synergizes malathion (**9**) in resistant houseflies by inhibiting the detoxifying malathion carboxylesterase and the GSH transferase that effects monodemethylation of this insecticide (Yeoh *et al.*, 1982). Phosphorothiolates such as S,S,S-tributylphosphorotrithioate, profenofos (**13**), methamidophos (**14**) and acephate (**15**), synergize the new synthetic pyrethroids in insects by inhibiting their detoxication by carboxylesterases. Combinations with pyrethroids are under intensive investigation against a variety of agricultural insect pests as a means to improve pyrethroid efficacy and to overcome resistance (Ozaki *et al.*, 1984; Ishaaya *et al.*, 1987).

In insects, carbamate insecticides are particularly susceptible to detoxification by MFO enzymes, rather than esterases, and the synergism seen with MFO inhibitors against both insecticide-susceptible and carbamate-resistant diptera is sometimes spectacular (Brooks, 1979). Simple derivatives of 1,2-methylenedioxybenzene (1,3-benzodioxole) and aryl 2-propynyl ethers, which inhibit insect cytochrome P-450, are the best-known insecticide synergists. Sulphur compounds with synergistic properties include 5,6-dichloro-1,2,3-benzothiadiazole and relatives, thanite (2-(2-thiocyanato)acetoxy-1,7,7-trimethylbicyclo[2.2.1]heptane), 4-nitrobenzylthiocyanate, and S,S,S-tributyl phosphorotrithioate. Benzothiadiazole derivatives are known to interact with cytochrome P-450; little is known of the action of the other compounds, although they may be potentially active against this cytochrome either directly or following conversion into reactive intermediates. The synergism of carbamate insecticides has not been exploited to any extent in the field, partly because of problems relating to co-formulation.

Current interest in pyrethroid synergism may stimulate renewed enthusiasm for synergistic combinations involving other insecticides. Such combinations can be selective in favour of mammals if carefully chosen but the possibility of enhanced toxicity to other non-target organisms must be kept in mind.

REFERENCES

Ashton, F. M. and Glenn, R. K. (1979). Influences of chloro-, methoxy-, and methylthio- substitutions of bis(isopropylamino)-s-triazine on selected metabolic processes. *Pestic. Biochem. Physiol.*, **11**, 201–207.

Balasubramanya, R. H., Patil, R. B., Bhat, M. V. and Nagendrappa, G. (1980). Degradation of carboxin and oxycarboxin by *Pseudomonas aeruginosa* isolated from soil. *J. Environ. Sci. Health*, **B15**, 485–505.

Bleeke, M. S., Smith, M. J. and Casida, J. E. (1985). Metabolism and toxicity of metribuzin in mouse liver. *Pestic. Biochem. Physiol.*, **23**, 123–130.

Brooks, G. T. (1974). In *Chlorinated Insecticides*, Vol. I, CRC Press, Cleveland, pp. 7–11.

Brooks, G. T. (1979). The metabolism of xenobiotics in insects. In J. W. Bridges and L. F. Chasseaud (eds.), *Progress in Drug Metabolism*, Vol. 3, Wiley, Chichester, pp. 151–214.

Brooks, G. T. (1984). Metabolism and pesticide design. In J. W. Bridges and L. F. Chasseaud (eds.), *Progress in Drug Metabolism*, Vol. 8, Taylor and Francis, London, pp. 101–202.

Brown, J. J. (1987). Toxicity of herbicides thiobencarb and endothall when fed to laboratory-reared *Trichoplusia ni*. *Pestic. Biochem. Physiol.*, **27**, 97–100.

Casida, J. E. and Ruzo, L. (1986). Reactive intermediates in pesticide metabolism: peracid oxidations as possible biomimetic models. *Xenobiotica*, **16**, 1003–1015.

Climie, I. J. G. and Hutson, D. H. (1979). Conjugation reactions with amino-acids including glutathione. In H. Geissbuhler, G. T. Brooks and P. C. Kearney, (eds.), *Advances in Pesticide Science*, Vol. 3, Pergamon, Oxford, pp537–546.

Colby, S. R., Barnes, J. W., Sampson, T. A., Shoham, J. L. and Osborn, D. J. (1983). Fomesafen — a new selective herbicide for post-emergence broadleaf weed control in soybean. In *Proc. 10th Int. Congr. Plant Protection* 1983, Vol. 1, British Crop Protection Council, Thornton Heath, pp. 295–302.

Cole, D. (1983). Oxidation of xenobiotics in plants. In D. H. Hutson and T. R. Roberts (eds.), *Progress in Pesticide Biochemistry and Toxicology*, Vol. 3, Wiley, Chichester, pp. 199–254.

De Baun, J. R., Miallus, J. B., Knarr, J., Mihailovski, A. and Menn, J. J. (1974). The fate of N-trichloro[^{14}C]methylthio-4-cyclohexene-1,2-dicarboximide (captan) in the rat. *Xenobiotica*, **4**, 101–119.

De Baun, J. R., Bova, D. L., Tseng, K. C. and Menn, J. J. (1978). Metabolism of [ring ^{14}C]Ordram (molinate) in the rat. 2. Urinary metabolite identification. *J. Agric. Food Chem.*, **26**, 1098–1104.

Dennis, S. and Kennedy, I. R. (1986). Monooxygenases from soybean root nodules: aldrin epoxidase and cinnamic acid 4-hydroxylase. *Pestic. Biochem. Physiol.*, **26**, 29–35.

Dodge, A. D. (1983). The mode of action of herbicides. In D. H. Hutson and T. R.

Roberts (eds.), *Progress in Pesticide Biochemistry and Toxicology*, Wiley, Chichester, pp. 163–197.

Drabek, J. and Neumann, R. (1985). Proinsecticides. In D. H. Hutson and T. R. Roberts (eds.), *Insecticides*, Wiley, Chichester, pp. 35–86.

Draber, W. and Fedtke, C. (1979). Herbicide interaction with plant biochemical systems. In H. Geissbuhler, G. T. Brooks and P. C. Kearney, (eds.), *Advances in Pesticide Science*, Pergamon, Oxford, pp. 475–486.

Dutton, F. E., Gemrich II, E. G., Lee, B. L., Nelson, S. J., Parham, P. H. and Seamen, W. J. (1981). Insecticidal phosphoramidothio-derivatives of the carbamate methomyl. *J. Agric. Food Chem.*, **29**, 1114–1118.

Engst, R. (1979). Chemical toxification of pesticides in the environment. In H. Geissbuhler, G. T. Brooks and P. C. Kearney, (eds.), *Advances in Pesticide Science*, Vol. 3, Pergamon, Oxford, pp. 590–597.

Esser, H. O. (1986). Comparative aspects of herbicide metabolism. *Xenobiotica*, **16**, 1031–1045.

Eto, M., Hashimoto, Y., Ozaki, K., Kassai, T. and Sasaki, Y. (1975). Fungitoxicity and insecticide synergism of monothioquinol phosphate esters and related compounds. *Botyu Kagaku*, **40**, 110–117.

Eya, B. K. and Fukuto, T. R. (1986). Formamidine S-carbamates: a new procarbamate analogue with improved ovicidal and acaricidal activities. *J. Agric. Food Chem.*, **34**, 942–952.

Fahmy, M. A. H., Mallipudi, N. M. and Fukuto, T. R. (1978). Selective toxicity of N,N'-thiodicarbamates. *J. Agric. Food Chem.*, **26**, 550–557.

Frear, D. S., Swanson, H. R. and Mansagar, E. R. (1985). Alternate pathways of metribuzin metabolism in soybeans: formation of N-glucoside and homoglutathione conjugates. *Pestic. Biochem. Physiol.*, **23**, 56–65.

Funayama, S., Uchida, M., Kanno, H. and Tsuchiya, K. (1986). Degradation of buprofezin in flooded and upland soils under laboratory conditions. *J. Pestic. Sci.*, **11**, 605–610.

Gorder, G. W., Kirino, O., Hirashima, A. and Casida, J. E. (1986). Bioactivation of isofenphos and analogues by oxidative N-dealkylation and desulphuration. *J. Agric. Food Chem.*, **34**, 941–947.

Green, G. H., McKeown, B. A. and Oloffs, P. C. (1984). Acephate in rainbow trout (*S. gairdneri*): acute toxicity, uptake, elimination. *J. Environ. Sci. Health*, **B19**, 355–377.

Harris, M., Price, R. N., Robinson, J. and May, T. E. (1986). WL108477 — a novel neurotoxic insecticide. In *Proc. 1986 British Crop Protection Conf. — Pests and Diseases*, British Crop Protection Council, Thornton Heath, pp. 115–122.

Hirashima, A., Leader, H., Holden, I. and Casida, J. E. (1984). Resolution and stereoselective action of sulprofos and related S-propyl phosphorothiolates. *J. Agric. Food Chem.*, **32**, 1302–1307.

Holloman, D. W. and Chamberlain, K. (1981). Hydropyrimidine fungicides inhibit adenosinedeaminase in barley powdery mildew. *Pestic. Biochem. Physiol.*, **16**, 158–169.

Horvath, L. and Pulay, A. (1980). Metabolism of EPTC in germinating corn: sulfone as the true carbamoylating agent. *Pestic. Biochem. Physiol.*, **14**, 265–270.

Hutchison, J. M., Shapiro, R. and Sweetser, P. B. (1984). Metabolism of chlorsulfuron by tolerant broadleaves. *Pestic. Biochem. Physiol.*, **22**, 243–247.

Ikeda, M., Unai, T. and Tomizawa, C. (1986a). Degradation of the herbicide orbencarb in soils. *J. Pestic. Sci.*, **11**, 85–96

Ikeda, M., Unai, T. and Tomizawa, C. (1986b). Absorption, translocation and metabolism of orbencarb in soybean plants. *J. Pestic. Sci.*, **11**, 97–110.

Imai, Y. and Kuwatsuka, S. (1982). Degradation of the herbicide molinate in soils. *J. Pestic. Sci.*, **7**, 487–497.

Imai, Y. and Kuwatsuka, S. (1984). Uptake, translocation, and metabolic fate of the herbicide molinate in plants. *J. Pestic. Sci.*, **9**, 79–90.

Imai, Y. and Kuwatsuka, S. (1986). Metabolic pathways of the herbicide molinate in four strains of isolated soil microorganisms. *J. Pestic. Sci.*, **11**, 245–251.

Ishaaya, I., Mendelson, Z., Ascher, K. R. S. and Casida, J. E. (1987). Cypermethrin synergism by pyrethroid esterase inhibitors in adults of the whitefly *Bemisia tabaci. Pestic. Biochem. Physiol.*, **28**, 155–162.

Kakiki, K. and Misato, T. (1979) Effect of isoprothiolane on fatty acid synthesis. *J. Pestic. Sci.*, **4**, 305–313.

Kazano, H., Koyama, S. and Sutrisno, (1983). Mechanisms of insecticidal selectivity of propaphos between the green rice leafhopper, *N. cincticeps* and the common cutworm, *S. Litura. J. Pestic. Sci.*, **8**, 561–565.

Kerkenaar, A. and Kars Sijpesteijn, A. (1977). On the mode of action of prothiocarb. *Neth. J. Plant Pathol.*, **83** (Suppl. 1), 145–152.

Kimura, S., Toeda, K., Miyamoto, T. and Yamamoto, I. (1984). Activation and detoxication of S-alkyl phosphorothiolate insecticides. *J. Pestic. Sci.*, **9**, 137–142.

Knowles, C. O. (1982). Structure–activity relationships among amidine acaricides and insecticides. In J. R. Coats (ed.), *Insecticide Mode of Action*, Academic Press, New York, pp. 243–277.

Kuhr, R. J. and Dorough, H. W. (1976). *Carbamate Insecticides*: *Chemistry, Biochemistry and Toxicology*, CRC Press, Cleveland.

Kurogochi, S., Katagiri, M., Takase, I. and Uesugi, Y. (1985). Metabolism of edifenphos by strains of *Pyricularia oryzae* with varied sensitivity to phosphorothiolate fungicides. *J. Pestic. Sci.*, **10**, 41–46.

Langcake, P., Kuhn, P. J. and Wade, M. (1983). The mode of action of systemic fungicides. In D. H. Hutson and T. R. Roberts (eds.), *Progress in Pesticide Biochemistry and Toxicology*, Vol. 3, Wiley, Chichester, pp. 1–109.

Lay, M. M. and Menn, J. J. (1987). Amelioration of toxicity to Japanese carp with selected dichloroacetamides. *Pestic. Biochem. Physiol.*, **28**, 149–154.

Lay, M. and Niland, A. M. (1985). Biochemical response of inbred and hybrid corn (*Zea mays* L.) to R-25788 and its distribution with EPTC in corn seedlings. *Pestic. Biochem. Physiol.*, **23**, 131–140.

Lee, P. W., Allahyari, R. and Fukuto, T. R. (1978). Studies on chiral isomers of fonofos and fonofos-oxon. *Pestic. Biochem. Physiol.*, **9**, 23–32.

Lee, P. W., Allahyari, R. and Fukuto, T. R. (1980). Absorption and metabolism of the chiral isomers of fonofos in the corn and cotton plant. *J. Environ. Sci. Health*, **B15**, 25–37.

Magee, T. A. (1982). Oxime carbamate insecticides. In J. R. Coats, (ed.), *Insecticide Mode of Action*, Academic Press, New York, pp. 71–100.

Marsden, P. J. and Casida, J. E. (1982). 2-Haloacrylic acids as indicators of mutagenic 2-haloacrolein intermediates in mammalian metabolism of selected promutagens and carcinogens. *J. Agric. Food Chem.*, **30**, 627–631.

Marsden, P. J., Kuwano, E. and Fukuto, T. R. (1982). Metabolism of carbosulfan in the rat and housefly. *Pestic. Biochem. Physiol.*, **18**, 38–48.

Mayer, P., Kriemsler, H. P., Hambock, H. and Laanio, T. L. (1981). Metabolism of O,O-dipropyl S-[2-(2'-methyl-1'-piperidinyl)-2-oxoethyl]phosphorodithioate (C19490) in paddy rice. *Agr. Biol. Chem.*, **45**, 355–360.

Menzie, C. (1980). *Metabolism of Pesticides, Update III*, United States Department of the Interior, Fish and Wildlife Service, Special Scientific Report — Wildlife No. 232, Washington, DC, p. 189.

Metcalf, R. A. and Metcalf, R. L. (1973). Selective toxicity of analogs of methyl parathion. *Pestic. Biochem. Physiol.*, **3**, 149–159.

Mine, A., Miyakado, M. and Matsunaka, S. (1975). The mechanisms of bentazon selectivity. *Pestic. Biochem. Physiol.*, **5**, 566–574.

Miyamoto, J. and Ohkawa, K. (1979). Oxidative processes in pesticide transformation. In H. Geissbuhler, G. T. Brooks and P. C. Kearney, (eds.), *Advances in Pesticide Science*, Vol. 3, Pergamon, Oxford, pp. 508–515.

Miyazaki, A., Nakamura, T. and Marumo, S. (1985). Chiral metabolism of achiral propaphos into R_S–(+)–propaphos sulfoxide in rice plants. *J. Pestic. Sci.*, **10**, 727–728.

Moon, Y. H. and Kuwatsuka, S. (1985). Microbial aspects of the herbicide benthiocarb in soil. *J. Pestic. Sci.*, **10**, 513–521.

Moon, Y. H. and Kuwatsuka, S. (1987). Population changes of benthiocarb dechlorinating microorganisms in soil. *J. Pestic. Sci.*, **12**, 11–16.

Muecke, W. (1983). Separation and purification of pesticide metabolites. In D. H. Hutson and T. R. Roberts (eds.), *Progress in Pesticide Biochemistry and Toxicology*, Vol. 3, Wiley, Chichester, pp. 279–366.

Muecke, W., Kriemler, H. P., Hug, P. and Alt, K. O. (1981). Metabolism of O,O-dipropyl S-[2-(2-methyl-1-piperidinyl)-2-oxo-ethyl]phosphorodithioate in the rat. *Agric. Biol. Chem.*, **45**, 43–51.

Mukerjee, S. K. (1982). Agrochemicals in India. In *Proc. Symp. on Agrochemicals*: *Fate in Food and the Environment, Rome, June, 7–11, 1982*, International Atomic Energy Agency, Vienna, pp. 3–21.

Nishi, K., Kodo, I. and Tan, N. (1979). Absorption and translocation of cartap in rice plant. *J. Pestic. Sci.*, **4**, 37–44.

Ohkawa, H. (1982). Stereoselectivity of organophosphorus insecticides. In J. R. Coats (ed.), *Insecticide Mode of Action*. Academic Press, New York, pp. 163–185.

Ohkawa, H., Shibaike, R., Okihara, Y., Morikawa, Y. and Miyamoto, J. (1976a). Degradation of the fungicide Denmert (S-1358) by plants, soils and light. *Agric. Biol. Chem.*, **40**, 943–951.

Ohkawa, H., Okihara, Y. and Miyamoto, J. (1976b) Metabolism of the fungicide Denmert and Denmert sulfoxides in liver enzyme systems. *Agr. Biol. Chem.*, **40**, 1175–1182.

Orr, G. L. and Hess, F. D. (1981). Characterization of herbicidal injury by acifluorfen-methyl in excised cucumber (*Cucumis sativus* L.) cotyledons. *Pestic. Biochem. Physiol.*, **16**, 171–178.

Otto, S., Bentel, P., Drescher, N. and Haber, R. (1979). Investigations into the degradation of bentazon in plant and soil. In H. Geissbuhler, G. T. Brooks and P. C. Kearney, (eds.), *Advances in Pesticide Science*, Pergamon, Oxford, pp. 551–556.

Ozaki, K., Sasaki, Y. and Kassai, T. (1984). The insecticidal activity of mixtures of pyrethroids and organophosphates or carbamates against the insecticide-resistant green rice leafhopper. *J. Pestic. Sci.*, **9**, 67–72.

Rosen, J. D., Magee, P. S. and Casida, J. E. (1981). *Sulphur in Pesticide Action and Metabolism*, ACS Symposium Series 158, American Chemical Society, Washington, DC.

Rosival, L., Vargova, M., Szokolayona, J., Cerey, K., Hladka, K., Batova, V., Kovacicova, J. and Truchlik, S. (1976). Contribution to the toxic action of S-methyl fenitrothion. *Pestic. Biochem. Physiol.*, **6**, 280–286.

Ryan, D. L. and Fukuto, T. R. (1985). The effect of impurities on the toxicokinetics of malathion in rats. *Pestic. Biochem. Physiol.*, **23**, 413–424.

Santi, R. and Gozzo, F. (1976). Degradation and metabolism of Drepamon in rice and barnyard grass. *J. Agric. Food Chem.*, **24**, 1229–1235.

Sattelle, D. B., Harrow, I. D., David, J. A., Pelhate, M., Callec, J. J., Gepner, J. I. and Hall, L. M. (1985). Nereistoxin: actions on a CNS acetylcholine receptor/ion channel in the cockroach, *Periplaneta americana*. *J. Exp. Biol.*, **118**, 37–52.

Schloss, J. V., Ciskanik, L. M., and Van Dyk, D. E. (1988). Origin of the herbicide binding site of acetolactate synthase. *Nature* **331,** 360–362.

Schuphan, I. and Casida, J. E. (1983). Metabolism and degradation of pesticides and xenobiotics: bioactivation involving sulphur containing substituents. In J. Miyamoto and P. C. Kearney (eds.), *Pesticide Chemistry*: *Human Welfare and Environment*, Vol. 3, Pergamon, Oxford, pp. 287–294.

Siegel, M. R. (1970). Reactions of certain trichloromethylsulfenyl fungicides with low molecular weight thiols. *J. Agric. Food Chem.*, **18**, 823–826.

Smyser, B. P., Sabourin, P. J. and Hodgson, E. (1985). Oxidation of pesticides by purified microsomal FAD-containing monooxygenases from mouse and pig liver. *Pestic. Biochem. Physiol.*, **24**, 368–374.

Somerville, L. (1986). The metabolism of fungicides. *Xenobiotica*, **16**, 1017–1030.

Thomas, V. M. and Holt, C. L. (1980). The degradation of [^{14}C]molinate in soil under flooded and non-flooded conditions. *J. Environ. Sci. Health*, **B15**, 475–484.

Uchida, M., Funayama, S. and Sugimoto, T. (1982). Bioconcentration of fungicidal dialkyl dithioanylidene malonates in *Oryzias latipes* L. *J. Pestic. Sci.*, **7**, 181–186.

Uchida, M., Sumida, M. and Hirano, A. (1986). Stereoselective metabolism of isoprothiolane and its sulfoxide in rat liver *in vitro* and in rice plants. *J. Pestic. Sci.*, **11**, 573–578.

Uchida, M., Izawa, Y. and Sugimoto, T. (1987). Inhibition of prostaglandin biosynthesis and oviposition by an insect growth regulator, buprofezin, in *N. lugens*. *Pestic. Biochem. Physiol.*, **27**, 71–75.

Ueji, M. and Tomizawa, C. (1986). Insect toxicity and anti-AChE activity of chiral isomers of isofenphos and its oxon. *J. Pestic. Sci.*, **11**, 447–451.

Umetsu, N. (1986). Studies on synthesis and metabolism of selective toxic derivatives of methylcarbamate insecticides. *J. Pestic. Sci.*, **11**, 493–503.

Umetsu, N., Tanaka, A. K. and Fukuto, T. R. (1985). Absorption, translocation and metabolism of the insecticide benfuracarb in plants. *J. Pestic. Sci.*, **10**, 501–511.

Usui, M. and Umetsu, N. (1986). Metabolism of the insecticide benfuracarb in the housefly. *J. Pestic. Sci.*, **11**, 401–408.

Veerasekaran, P., Kirkwood, R. C. and Parnell, E. W. (1981). Studies of the mechanism of action of asulam in plants. Part II: effect of asulam on the biosynthesis of folic acid. *Pestic. Sci.*, **12**, 330–338.

Vilanova, E., Johnson, M. K. and Vicedo, J. L. (1987). Interaction of some unsubstituted phosphoramidate analogs of methamidophos (O,S-dimethyl phosphoramidate) with acetylcholinesterase and neuropathy target esterase of hen brain. *Pestic. Biochem. Physiol.*, **28**, 224–238.

Vonk, J. W. (1983). Metabolism of fungicides in plants. In D. H. Hutson and T. R. Roberts (eds.), *Progress in Pesticide Biochemistry and Toxicology*, Vol. 3, Wiley, Chichester, pp. 111–162.

Wilkinson, R. E. (1986) Diallate inhibition of gibberellin biosynthesis in sorghum coleoptiles. *Pestic. Biochem. Physiol.*, **25**, 93–97.

Wing, K. D., Glickman, A. H. and Casida, J. E. (1984). Phosphorothiolate pesticides and related compounds: oxidative bioactivation and aging of the inhibited acetylcholinesterase. *Pestic. Biochem. Physiol.*, **21**, 22–30.

Woloshuk, C. P. and Sisler, H. D. (1982). Tricyclazole, pyroquilon, tetrachlorophthalide, PCBA, courmarin and related compounds inhibit melanization and epidermal penetration of *P. oryzae*. *J. Pestic. Sci.*, **7**, 161–166.

World Health Organisation (1985). *Data Sheet on Pesticides No. 71. Thiram.*

Worthing, C. R. and Walker, S. B. (eds.) (1987). *The Pesticide Manual*, 8th edn., British Crop Protection Council, Thornton Heath.

Yamaguchi, I. (1982). Fungicides for control of rice-blast disease. *J. Pestic. Sci.*, **7**, 307–316.

Yeoh, C. L., Kuwano, E. and Eto, E. (1982). Effect of the fungicide IBP as a synergist in the metabolism of malathion in insects. *J. Pestic. Sci.*, **7**, 31–40.

4

Sulphur compounds as industrial and medicinal agents

L. A. Damani
Chelsea Department of Pharmacy, Kings College London, Manresa Road, London, SW3 6LX, UK, and
M. Mitchard
Medical Directorate, Glaxo Group Research Ltd., Greenford, Middlesex, UB6 0HE, UK

SUMMARY

1. Elemental sulphur and many of its organic and inorganic derivatives have proved to be useful in a number of industrial, agricultural and medicinal applications.
2. The main agricultural use of elemental sulphur is in the synthesis of sulphuric acid for the subsequent manufacture of fertilizers. Other agricultural uses include the use of organosulphur compounds as insecticides, herbicides and fungicides.
3. The principal non-agricultural industrial uses of sulphur include the manufacture of paper, fabrics, rubber, petroleum, steel, dyes, paints and thousands of other household products. Organic solvents such as carbon disulphide and dimethylsulphoxide have found many industrial applications.
4. Sulphur is also used in medicine. There is hardly a class of drugs which does not contain compounds having sulphur in their structure. Organosulphur compounds form a major therapeutic resource and their potential remains to be fully exploited.

4.1 INTRODUCTION

Detailed descriptions of the naturally occurring sulphur compounds and of the agricultural importance of synthetic sulphur xenobiotics are given in chapters 2 and 3 of this volume. It is the purpose of this chapter to survey the industrial and medicinal uses of elemental sulphur, and its inorganic and organic derivatives. Together, these three chapters should provide the proper orientation to the subject of 'sulphur xenobiochemistry' expounded in subsequent chapters of this volume, and indeed should provide the *raison d'être* for the increased interest in this field in the last decade.

Although elemental sulphur has been known and used since antiquity, it is still of vital importance to the chemical industry today. It was the development of efficient processes for manufacturing sulphuric acid that made sulphur the workhorse of the early chemical industry (Fike, 1972). A large proportion ($\approx$85%) of mined sulphur is still converted to sulphuric acid, which is subsequently used in the manufacture of fertilizers and other chemicals and in the wood pulp industry. The majority of the rest is used as 'non-acid sulphur' in the manufacture of carbon disulphide and miscellaneous other chemicals. The remainder, a very small percentage, is used in the refined elemental form. In 1985, the global use of sulphur amounted to 57 million tons, with agricultural use accounting for 65% and non-agricultural use for 35% of sulphur consumption (The Sulphur Institute, 1986–1987).

Traditionally, the largest end-use of sulphur has been in the manufacture of fertilizers which supply essential soil nutrients and hence improve crop yield. In view of the increasing occurrence of sulphur deficiencies in soils in various parts of the world, there is likely to be a significant growth in the demand for sulphur-based fertilizers. Elemental sulphur itself and various organosulphur compounds are also used as fungicides, insecticides and herbicides (see chapter 2, this volume). Although non-agricultural use of sulphur is lower than agricultural use at present, this is set to change in the very near future because various new industrial applications have been developed which utilize the unique chemical and physical properties of this element. These new applications include acid leaching of copper from ores, use of emulsified sulphur in pulpmills and the use of corrosion-resistant sulphur-concretes and sulphur-enhanced asphalt as novel construction materials. Finally, organosulphur compounds are now widely used in medicine although, in terms of total sulphur use, this accounts for a relatively small percentage. However, the intentional exposure of man to organosulphur compounds, and the incidental exposure due to the use of sulphur in industry and agriculture, makes it essential to have a clear understanding of the biological fate of such compounds in living systems.

4.2 SULPHUR AND SULPHUR COMPOUNDS AS AGRICULTURAL AGENTS

Sulphur is a relatively low cost material, widely available in pure form. Its exceptional chemical versatility has resulted in an increase in use which has outstripped that of many other minerals and chemicals over the last two decades (Fig. 1(a)). High-sulphur containing phosphate fertilizers have been developed as a means of providing plant nutrient sulphur and consequently its use in agriculture has grown considerably during this period (Fig. 1(b)). Currently, sulphur is used principally in the form of inorganic and organic compounds. However, other industrial applications are being developed which exploit native sulphur's unique physical and chemical properties (see section 4.3.4).

4.2.1 Sulphur fertilizers
Sulphur is an essential plant nutrient, required in amounts similar to that for phosphorus. The cycling of sulphur in nature occurs in a manner similar to that of

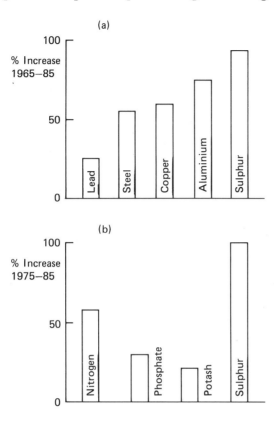

Fig. 1 — (a) Percentage growth in some mineral and chemical industries during 1965–1985; (b) percentage world growth in use of plant nutrients during 1975–1985. Data compiled from The Sulphur Institute (1986–1987).

nitrogen: a number of factors are therefore involved in the maintenance of adequate sulphur levels in the soil (Fig. 2). Loss of sulphur occurs primarily by removal of crops and through leaching. Sulphur is added to soils in the form of sulphur-based fertilizers and animal excreta and from rainfall containing dissolved atmospheric sulphur dioxide. Plants utilize sulphur as the sulphate for the biosynthesis of sulphur-containing amino acids upon which animals depend for their requirements of these essential amino acids. Animals in turn oxidize these amino acids to sulphates which are excreted in urine and faeces (Anderson, 1978).

In many parts of the world, soil nutrients are being exhausted rapidly as a result of increasing crop production. In the past, commonly used fertilizers, such as ammonium sulphate and ordinary 'superphosphate', contained considerable amounts of sulphur which met the crop requirements. Sulphuric acid is used in the manufacture of ammonium sulphate by direct reaction with gaseous ammonia:

$$2NH_3 + H_2SO_4 \rightarrow (NH_4)_2SO_4 \qquad (1)$$

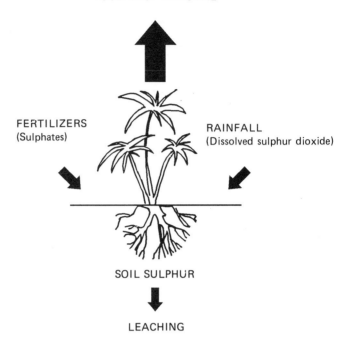

FOOD CROP REMOVAL

FERTILIZERS
(Sulphates)

RAINFALL
(Dissolved sulphur dioxide)

SOIL SULPHUR

LEACHING

Fig. 2 — Sulphur balance in agricultural fields.

Superphosphate is the product formed when sulphuric acid reacts with insoluble calcium phosphate, to give a soluble product that can be used as an effective fertilizer (see eq. (2)). Superphosphate, although regarded as a phosphate fertilizer, contains large amounts of sulphate.

$$Ca_3(PO_4)_2 + 2H_2SO_4 \rightarrow Ca(H_2PO_4)_2 + 2CaSO_4 \tag{2}$$

The increased demand for the essential nutrients, nitrogen, phosphorous and potassium, associated with the high costs of bulk transportation, have led to the use of high-analysis fertilizers (Braun, 1988). Since these fertilizers contain little or no sulphur, sulphur deficiencies are now being reported in many areas of the world. Ironically, a further factor in soil sulphur reduction may be the decrease in atmospheric sulphur dioxide produced as a result of pollution control legislation. Clearly, much of the sulphur lost from soil will have to be replaced, since sulphur deficiencies will otherwise severly limit crop production.

The bulk of the sulphur-containing fertilizers contain sulphate as the sulphur source; this constitutes the major plant-useable form of the element. There is, however, currently an interest in the use of reduced forms of sulphur (e.g. elemental sulphur, thiosulphate salts and polysulphides) as additives to soil fertilizers; these provide a controlled release of sulphate as the compounds are gradually oxidized in the soil. Thiosulphate ($S_2O_3^{2-}$) is readily oxidized in neutral or alkaline soil to sulphate, with tetrathionate as the intermediate oxidation state (Nor and Tabatabai, 1976):

$$2S_2O_3^{2-} \rightarrow S_4O_6^{2-} \rightarrow 4SO_4^{2-} \tag{3}$$

In acid soils, thiosulphate undergoes spontaneous decomposition to form elemental sulphur and either sulphate or sulphur dioxide (Tisdale *et al.*, 1985). Elemental sulphur is of interest as a fertilizer because of the favourable shipping and handling costs associated with this 'sulphur concentrate'. Chemolithotrophic bacteria of the genus *Thiobacillus* are the most commonly found sulphur-oxidizing microorganisms in the soil, although soil fungi and several heterotropic bacteria can also contribute to this oxidation (Wainwright, 1984). The oxidation of sulphur (see (4)) in the soil is influenced by a variety of factors, e.g. particle size, bacterial population, soil type, temperature, companion nutrients and methods of application to the soil.

$$S^0 \rightarrow S_2O_3^{2-} \rightarrow SO_4^{2-} \tag{4}$$

A clear understanding of these factors is urgently required, since this will allow the development of sulphur-containing fertilizers that slowly release soluble sulphate, thereby ensuring efficient use of the applied sulphur.

4.2.2 Sulphur pesticides

Elemental sulphur has been used for many years as a non-systemic, direct-acting fungicide, insecticide and acaricide. Indeed, the use of elemental sulphur as a fertilizer (see section 4.2.1 above) developed from its original use as a fungicide. Sulphur is applied to foliage as an ultrafine aqueous suspension (<10 microns in diameter). Although there is some evidence that the beneficial effects of native sulphur result from its fungicidal activity, a large proportion of the sulphur applied is apparently not taken up by the leaves. The unabsorbed sulphur washes into the soil, where the fine particles are oxidized very rapidly to sulphate (Boswell, 1988).

Apart from elemental sulphur, two other well-known sulphur compounds, calcium polysulphide and carbon disulphide, are used in pest control. Calcium polysulphide (CaS_x), or lime sulphur, readily decomposes under slightly acidic conditions to give elemental sulphur and hydrogen sulphide. Carbon disulphide has been widely used as a fumigant for the protection of stored grain from insect pests. This simple fumigant can also be applied as the chemical precursor, dazomet; this liberates carbon disulphide on acid hydrolysis (see chapter 2, this volume).

In addition to the uses of simple sulphur compounds as pesticides, a large number of very sophisticated organosulphur compounds have been designed for use as insecticides, acaricides, nematocides, herbicides and fungicides. More than one-third of the synthetic organic pesticides in use today contain sulphur (Worthing and Walker, 1987). The introduction of sulphur-containing moieties such as thioether, thione, thiol or heterocycle has led to the development of pesticides with enhanced selectivity. In addition, the metabolic conversion of the sulphur moieties can alter the toxicokinetics of pesticides; in some cases metabolism leads to selective activation in the pest, and deactivation in the host or economic species. Thus, the introduction of sulphur-containing moieties into pesticides is sometimes accompanied by a reduction in toxicity to mammals (Rosen *et al.*, 1981).

Most sulphur-containing pesticides are activated by metabolic oxidation at the sulphur atom. The phosphorothionate insecticides can be cited as an example, since compounds such as parathion (**1**) and malathion (**2**) undergo oxidative desulphuration to the corresponding active oxo-analogous. In these compounds, metabolic oxidation produces an increase in the reactivity of these insecticides as phosphorylating agents (Hutson and Logan, 1983). Thiocarbamate herbicides, such as EPTC (**3**), and alkylthiotriazine herbicides such as cyanatryn (**4**), provide further examples of this selective bioactivation process. EPTC is metabolically converted to the thiocarbamate sulphoxide, an active carbamoylating agent. Cyanantryn is also metabolized to the sulphoxide, a powerful triazinylating agent which shows a particular reactivity with thiols. A further description of sulphur-containing pesticides is given in chapter 3, in this volume.

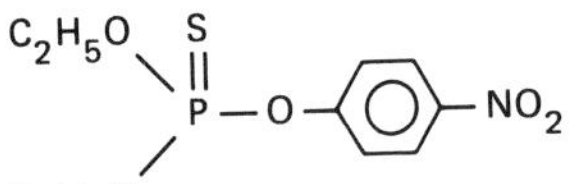

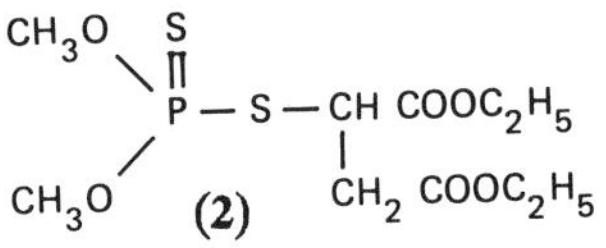

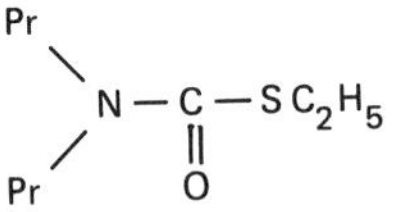

4.3 SULPHUR AND SULPHUR COMPOUNDS AS INDUSTRIAL AGENTS

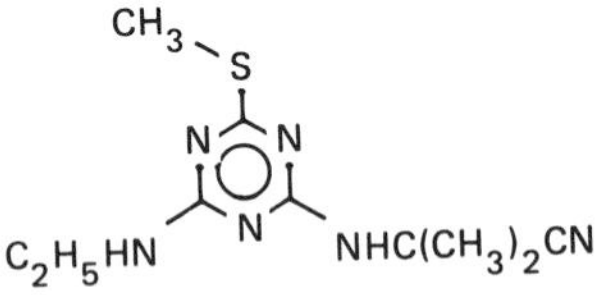

Elemental sulphur and many of its organic and inorganic derivatives are used in a number of industrial processes. These include the vulcanization of rubber, the manufacture of wood pulp, synthetic fibres (acrylic, polyester and rayon), detergents, dyes, plastics and cellophane, the processing of steel and, more recently, the leaching of copper ores and the manufacture of sulphur cements. This extensive industrial use produces an increase in the exposure of organisms and animals to products containing sulphur. The 'non-sulphur' industries that consume coal, natural gas and petroleum also contribute to the release of sulphur as oxides into the biosphere. The study of the fate of these sulphur compounds in various biological systems may help us to understand, and so to control, their damaging effects.

4.3.1 Sulphur in the rubber industry

Sulphur has a very important non-agricultural use in the rubber industry. Natural rubber is a mixture of linear hydrocarbon polymers, containing isoprene ($CH_2=C(CH_3)CH=CH_2$) as the repeating unit. The number of monomer units per molecule may vary from 1000 to 5000, giving molecular weights for the hydrocarbons from 60 000 to 350 000. *Rubber latex*, the raw material obtained from the rubber tree, is a colloidal suspension of rubber in water. Treatment of this material with acetic acid and salts affords *gum rubber*, a substance with elastic properties but with little tensile strength. Commercial exploitation of this material only became possible after a chance discovery by Charles Goodyear in 1839. He discovered that addition of sulphur to hot rubber causes changes that considerably improve the physical properties of rubber; this became the *vulcanization process* (Fieser and Fieser, 1963). The sulphur creates cross-links between the linear polymer chains of rubber, probably through saturation of the double bonds (Porter, 1968):

$$-(CH_2C(CH_3)=CHCH_2)-_n \qquad -(CH_2C(CH_3)=CHCH)-_n$$
$$+S \rightarrow \qquad | \qquad (5)$$
$$-(CH_2C(CH_3)=CHCH_2)-_n \qquad -(CH_2C(CH_3)-CHCH_2)-_n$$
$$|$$
$$S$$
$$|$$

Soft rubber contains approximately 2% sulphur, whereas *hard rubber* contains up to 35% sulphur. The latter material has lost the plasticity of natural rubber and contains fully saturated polymer chains that are extensively cross-linked. This use of sulphur has declined in recent years because of the development of synthetic elastomers that have properties superior to those of natural rubber.

4.3.2 Sulphur in the paper industry

Another important use of sulphur is in the large-scale manufacture of wood pulp used for the production of paper. Wood contains *cellulose*, a high molecular weight glucose polymer, and *lignin*, a non-carbohydrate complex phenolic polymer. Traditionally two alternative processes have been employed for the conversion of wood into paper, both designed to remove the undesired lignin as water-soluble derivatives. The *sulphite process* entails the digestion of wood using steam and an alkaline solution of bisulphites ($Mg(HSO_3)_2$; $Ca(HSO_3)_2$) at high pressure. This leads to the sulphonation of the phenolic aromatic nuclei in lignin, thus making it soluble. In the *alkali process*, lignin is degraded and solubilized by virtue of the acidity of the phenolic hydroxyl groups. Therefore, both processes remove most of the lignin and various other unwanted compounds from the wood, leaving the cellulose (pulp) ,behind. The 'sulphite waste liquor', containing excess bisulphites and sulphonated lignin, is invariably discharged into a river and constitutes a major source of pollution of the water supplies in paper-producing countries. A newly developed process of thermomechanical pulping should decrease this type of pollution hazard.

Sulphur in the rayon and cellophane industries

Although a large proportion of cellulose is used in the manufacture of paper, substantial amounts are also chemically modified to produce the polymeric materials, rayon and cellophane, that are familiar commercial products. The chemistry underpinning this industry dates back to when the organic chemistry of sulphur really commenced with the discovery of the *xanthate reaction* by Zeise in 1815:

$$R-OH + KOH + CS_2 \rightarrow R-O-\overset{\displaystyle S}{\overset{\|}{C}}-S^-K^+ + H_2O \qquad (6)$$

$$R-O-\overset{\displaystyle S}{\overset{\|}{C}}-S^-K^+ \xrightarrow{\text{acid}} R-O-\overset{\displaystyle S}{\overset{\|}{C}}-SH \qquad (7)$$

$$R-O-\overset{\displaystyle S}{\overset{\|}{C}}-SH \rightarrow R-OH + CS_2 \qquad (8)$$

Xanthation is a general reaction of alcohols which can occur at the alcoholic groups that occur along the spine of the cellulose molecule (O'Shaugnessy, 1958). The process of xanthation essentially brings the cellulose into solution as xanthates (eq. (6)). These may be re-precipitated in acid under appropriate conditions to afford fibres (rayon) or film (cellophane) with the re-formation of the initial reactant, carbon disulphide eq.s (7) and (8)). For cellophane production, the solubilized cellulose or *viscose*, is extruded as a sheet into an acid bath from which it is passed through several washings and a dryer to be wound on a roll: in the case of rayon, the viscose is extruded through a spinnaret that has 80 or more tiny holes, directly into an acid bath. The filaments unite to form a thread which is washed and wound onto a cone (Reid, 1963). More modern processes use acetic anhydride instead of acid to produce cellulose acetate which can also be spun into fibre.

4.3.4 Sulphur in other miscellaneous industries

The use of carbon disulphide as a pesticide was mentioned in section 4.2.2 and in the rayon and cellophane industries in section 4.3.3. In addition, this simple compound ranks with ethanol and benzene as the starting material for the synthesis of a large number of organic compounds. Those derived from carbon disulphide include dithioacids and esters, thiocyanates, dithiocarbonates and dithiocarbamates. Carbon disulphide is one of the most widely used organosulphur solvents. Dimethylsulphoxide is also used extensively in industry for its unique solvent properties and in medicine for its ability to promote the absorption of various pharmaceutical products through the skin. Sulphur occurs in many of the products we use in our daily lives such as dyes, paints and detergents. It is also used in the processing and manufacturing of steel, copper and other metals.

As can be seen, most uses of sulphur employ compounds containing the element. However, elemental sulphur does possess some unusual physical and chemical properites, and new uses are beginning to emerge. One application is in the manufacture of corrosion-resistant sulphur concrete. Asphaltic building materials deteriorate in time, through photochemical, chemical and biochemical (bacterial) action (Traxler, 1961). Addition of sulphur (20–25%) to asphalt produces sulphurized asphalt, which is more resistant to weathering and more durable than normal asphalt, especially at lower temperatures. Potential applications of appropriately modified sulphur are described in two excellent reviews by Barnes (1965) and Fike (1972).

4.4   SULPHUR AND SULPHUR COMPOUNDS AS MEDICINAL AGENTS

Elemental sulphur has long been used as a medicinal agent. It has been used for its scabicidal, insecticidal, fungicidal and purgative properties for many centuries. Its use as a mild laxative is still remembered by older people who, as children, were given 'brimstone and treacle' to keep them 'regular'. The mechanism of action of sulphur as a laxative, and indeed as a pesticide, is not clearly understood. It is most likely that its action is due to conversion to hydrogen sulphide and other toxic sulphides in biological systems (see Chapter 1, Volume 3, Part A, of this series). The use of sulphur for its purgative and fungicidal properties is now decreasing, although products containing sulphur, e.g. sulphur lozenges and sulphur-containing ointments and creams, are still available in pharmacies.

In the last two decades an ever-increasing number of organosulphur compounds have been introduced into medicine. These compounds contain sulphur in a variety of forms and oxidative states, such as thioethers, sulphoxides, sulphones, thiols and thiones. Sulphur-containing medicinal agents are represented in almost all therapeutic classes (Table 1), yet the introduction of sulphur in medicine was largely accidental. The early compounds were developed initially as dyes. Of particular interest are the sulphonic acid dyes, trypan blue and trypan red, the azo dye, prontosil and the phenothiazine, methylene blue. From these, the first effective antimicrobial and antiprotozoal agents were discovered. The early compounds paved the way for the therapeutic revolution of this century, with literally thousands of medicinal agents being developed from the original few compounds. In more recent years, sulphur has been deliberately introduced into medicinal compounds, to improve bioavailability or pharmacokinetic properties (Damani, 1987). Other sulphur compounds used in medicine have been developed from naturally occurring molecules.

The elemental composition of drugs in clinical use is of interest. From a sample of a thousand substances examined, 91% contained nitrogen, 25% sulphur, and 62% a heterocycle. Of these heterocycles, 95% again contained nitrogen, 24% sulphur and

Table 1 — Examples of some clinically used sulphur-containing drugs

Sulphur moiety	Therapeutic class	Examples of drugs in each therapeutic class
Thioether $(R-S-R^1)$	Antipsychotics, CNS drugs	Phenothiazines (e.g. chlorpromazine, dothiepin)
	Histamine H_2-receptor antagonists	Cimetidine, ranitidine
	Mucolytic	Carbocisteine
	Anthelmintics	Thiabendazole, phenothiazine
	Hypnotics	Chlormethiazole
Sulphoxide $\begin{array}{c} O \\ \parallel \\ (R-S-R^1) \end{array}$	Anti-rheumatic	Sulindac
	Anti-platelet aggregant	Sulphinpyrazone
	Uricosuric	Sulphinpyrazone
Sulphone $\begin{array}{c} O \\ \parallel \\ (R-S-R^1) \\ \parallel \\ O \end{array}$	Antimicrobial	Dapsone
Sulphonamide $(R-SO_2NH_2)$	Antimicrobial	Sulphadimidine (and other antibacterial sulphonamides)
Thiol $(R-SH)$	Antihypertensive	Captopril
	Cytotoxic agents	Thioguanine, mecaptopurine
	Chelating agent	Penicillamine
Disulphide $(R-S-S-R^1)$	Treatment of alcoholism	Disulfiram
Thione $(>C=)$ or $\geqslant P=S)$	Intravenous anaesthetics	Thiopentone
	Antithyroid agents	Carbimazole, propylthiouracil
	Antitubercular drugs	Ethionamide
	Anti-infectives	Malathion, pyrithione

16.5% oxygen as the heteroatom. Since the figures exceed 100%, it is clear that many compounds contain not only nitrogen but also sulphur and oxygen in the molecule. Sulphur heterocycles commonly found in drug molecules include phenothiazine, thiophene, thiazole, thiazolidine and others. These interesting statistics from Roth and Kleemann (1988) clearly indicate the importance of sulphur in medicine.

4.4.1   Sulphur-containing drugs developed from dyes

The early work in Germany on antimicrobial chemotherapy with various dyes led eventually to development of many of today's drugs. Paul Ehrlich's earlier success with trypan red led to the synthesis of several related naphthalene sulphonic acids which culminated in the discovery of suramin (**5**) for the treatment of trypanosomiasis (Klein, 1966). In this class of compounds, the sulphonic acid residues are essential for trypanocidal activity. Although methylene blue (**6**) was successful as an antimalarial agent, subsequent work with this class of chemicals (i.e. phenothiazines) led to the development of drugs for other uses; for example the antihistamines (prometha-

zine, **7**), antipsychotic phenothiazines (chlorpromazine, **8**) and antipsychotic dibenz-
thiapines (dothiepin, **9**) (Biel *et al.*, 1978).

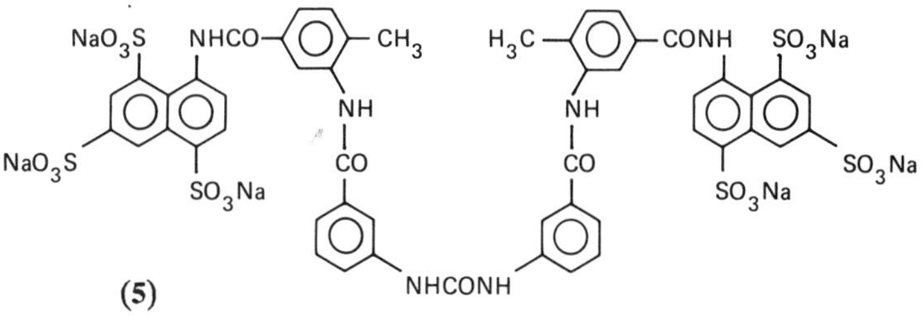

The most spectacular of the drug development programmes started with the azo
dye prontosil (**10**). Domagk discovered the antibacterial activity of this compound in
mice in 1932 and subsequently in man (Otten, 1986). Three years later he demonstrated that this activity was due to the metabolite sulphanilamide. A large number of
'sulphonamides' have been synthesized from this first lead compound. Some compounds have been designed for the treatment of gastrointestinal infections and are
not absorbed to any extent. Others were developed specifically for the treatment of
topical (e.g. corneal) infections. Yet others were designed as systemic anti-bacterial
agents, some with very short half-lives for urinary tract infections, and others with
long half-lives for systemic infections.

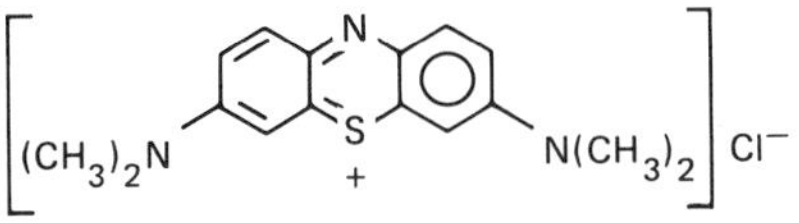

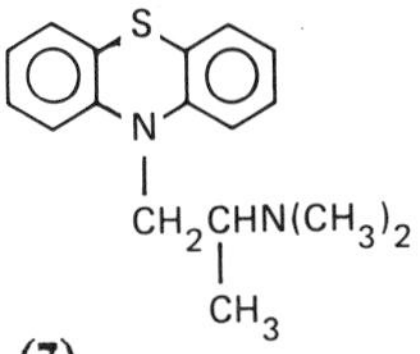

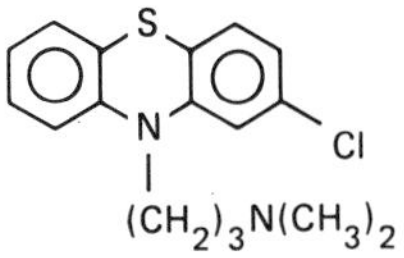

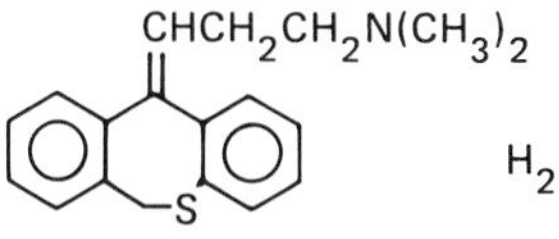

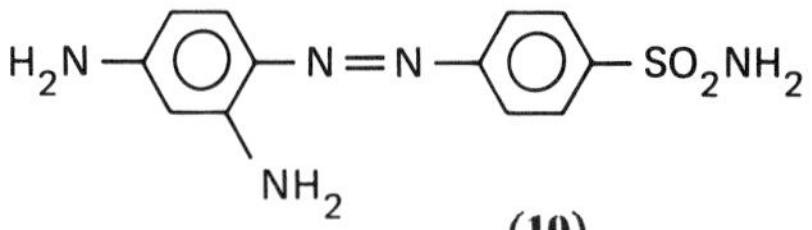

The widespread use of sulphonamides in man led to some important pharmacological discoveries. A large number of drugs were subsequently developed with widely
different therapeutic activities, in most of which the benzenesulphonamide moiety is
recognizably retained. These classes of drugs include diuretics (acetazolamide, **11**),
uricosuric agents (probenecid, **12**), antileprotics (dapsone, **13**) oral hypoglycaemic

agents (chlorpropamide, **14**) and anti-thyroid drugs (methimazole, **15**). Fig. 3 shows the 'chemical tree' of the sulphonamide drugs, from which it can be seen that only two are used as antibacterial agents. The diversity of the drugs developed from the sulphonamide structure isolated from prontosil is, to say the least, impressive. The process of 'molecular roulette', so often denigrated, has in this case produced great rewards.

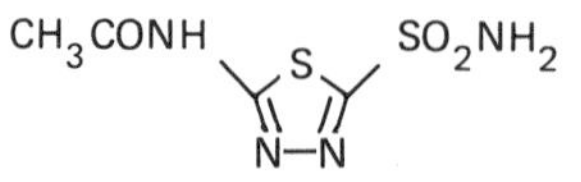

(11)

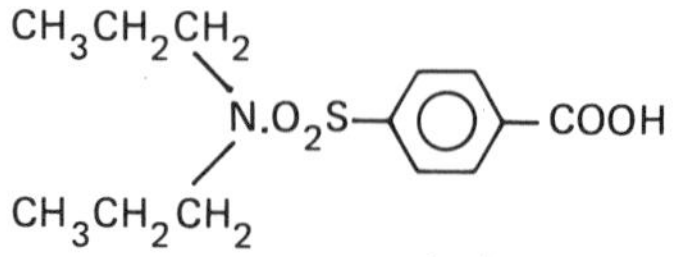

(12)

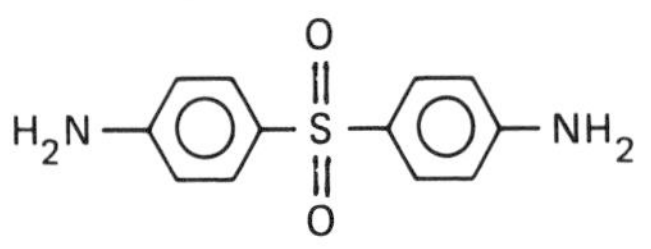

(13)

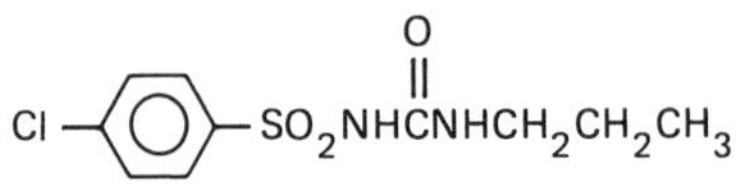

(14)

(15)

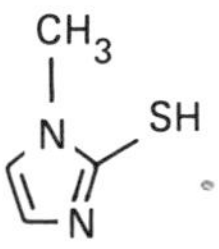

4.4.2 Sulphur-containing drugs developed from naturally occurring compounds

The major group of therapeutic agents in this class is the β-lactam antibiotics. The penicillins and cephalosporins have a common biosynthetic pathway; both are synthesized from a common tripeptide (δ(-α-aminoadipyl)-L-cysteinyl-D-valine (see Fig. 4). The precursor tripeptide gives rise to 6-aminopenicillanic acid, which contains the four-membered β-lactam ring fused to a thiazolidine ring. In some microorganisms, the thiazoline ring re-arranges to a six-membered dihydrothiazine ring, to give compounds based on the 7-aminocephalosporanic acid. This re-arrangement can also be effected chemically, by initially oxidizing penicillin to its sulphoxide (Cooper and Spry, 1971). The biosynthetic β-lactam antibiotics (e.g. benzylpenicillin, cephalosporin C) have been modified by semi-synthetic processes to a variety of compounds with a broad spectrum of antibacterial activity (Neu, 1985; Newall, 1985).

It had previously been supposed that the β-lactam ring had to be fused to one of the two sulphur heterocycles (thiazolidine or dihydrothiazine ring) for antibacterial activity. This is now known not to be so. Thienamycin (**16**) was isolated from *Streptomyces cattleya* and although this is too unstable to be used as an antibiotic, the *N*-formimidoyl derivative is stable and has been developed as imipenem (**17**) (Clissold *et al.*, 1987). The final group in the β-lactams are the monobactams or monocyclic β-lactam antibiotics. The monobactum nucleus is the sulphamate 3-aminomonobactamic acid (3-AMA, **18**). Aztreonam (**19**) is one of the amino-substituted derivatives recently used in clinical practice (Bonner and Sykes, 1984).

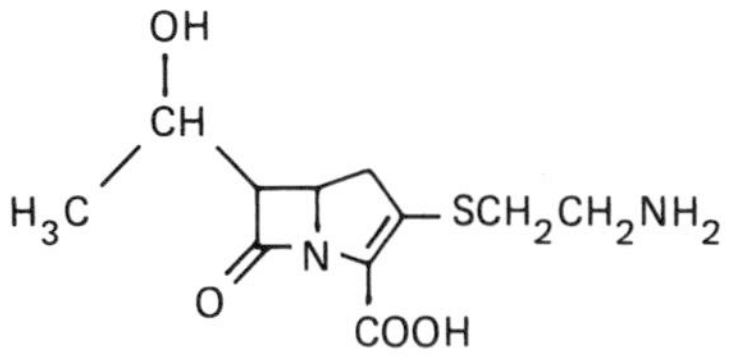

(**16**) 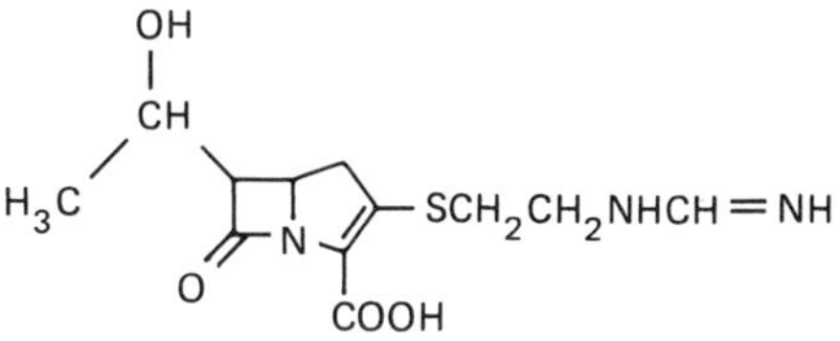   (**17**)

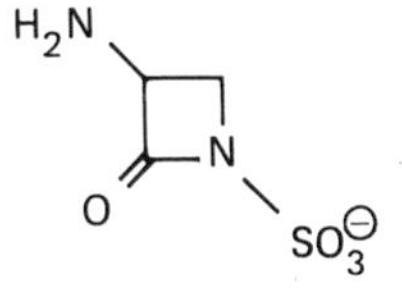

(**18**) 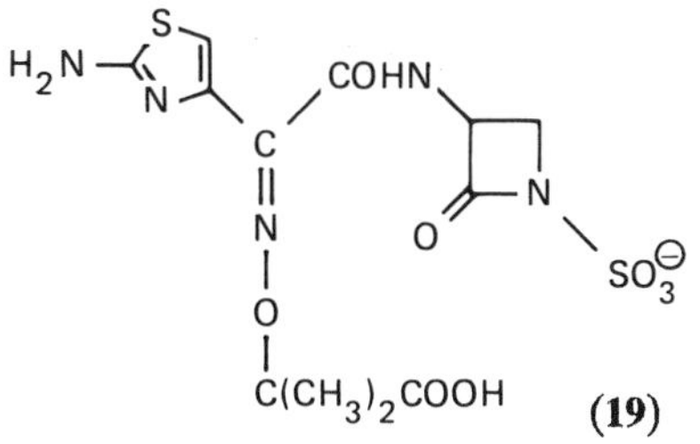   (**19**)

4.4.3 Drugs synthesized specifically to include sulphur

A number of drugs have been synthesized in recent years specifically to include sulphur. These include compounds in which the oxygen molecule has been replaced by the isosteric sulphur atom, and those molecules in which sulphur has been introduced to modify the electric field potential and thereby to 'shape' the molecule to produce the desired effect. In many cases the role of the sulphur functionality in drug action, or in drug disposition and toxicity, was only elucidated in retrospective studies. Sulindac (**20**) is an interesting example of a drug developed as a result of the search for a non-steroidal anti-inflammatory agent less toxic than indomethacin (**21**). Sulindac, a sulphoxide, is a prodrug which undergoes reduction *in vivo* to an active sulphide (Shen and Winter, 1977). Sulphinpyrazone (**22**) is another example of a sulphoxide drug that was synthsized in an effort to improve the uricosuric properties of phenylbutazone (**23**). In this case sulphoxide reduction decreases the uricosuric activity of the drug. Sulphinpyrazone has also recently been evaluated as an anti-thrombotic agent because of its ability to prevent platelet aggregation. Sulphoxide reduction in this case results in an increase in pharmacological activity, the sulphide

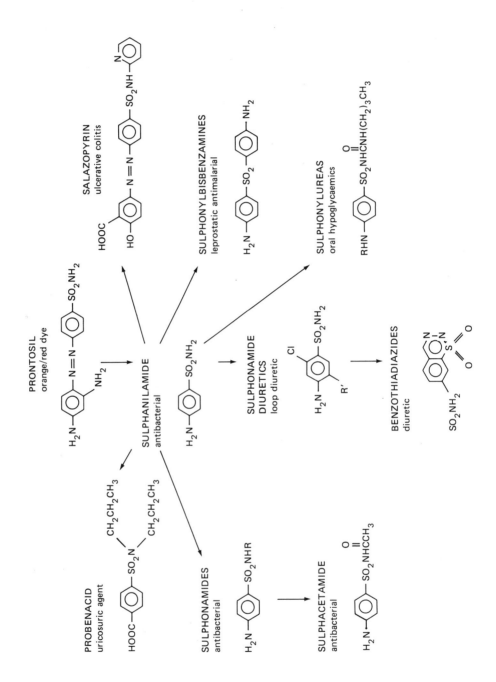

Fig. 3 — 'Chemical tree' of sulphonamide drugs.

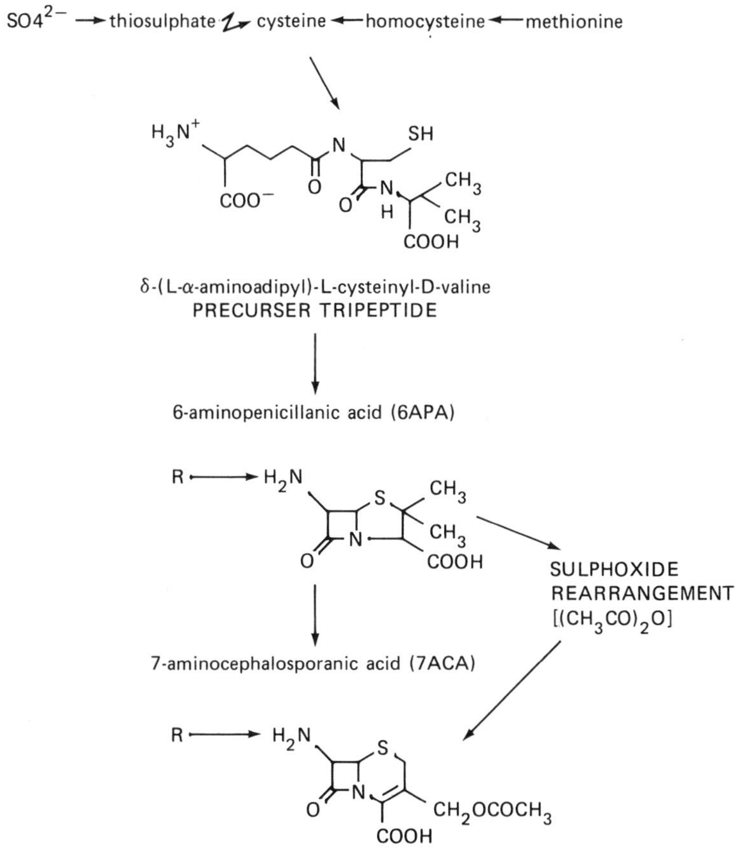
SO4[2−] → thiosulphate → cysteine ← homocysteine ← methionine
H_3N^+
COO−
SH
CH_3
COOH
δ-(L-α-aminoadipyl)-L-cysteinyl-D-valine
PRECURSER TRIPEPTIDE
6-aminopenicillanic acid (6APA)
R → H_2N
SULPHOXIDE
REARRANGEMENT
[$(CH_3CO)_2O$]
7-aminocephalosporanic acid (7ACA)
CH_2OCOCH_3

Fig. 4 — Biosynthetic pathways of the β-lactam antibiotics.

being a more potent anti-platelet aggregation agent (see Chapter 5, Volume 1, part B of this series).

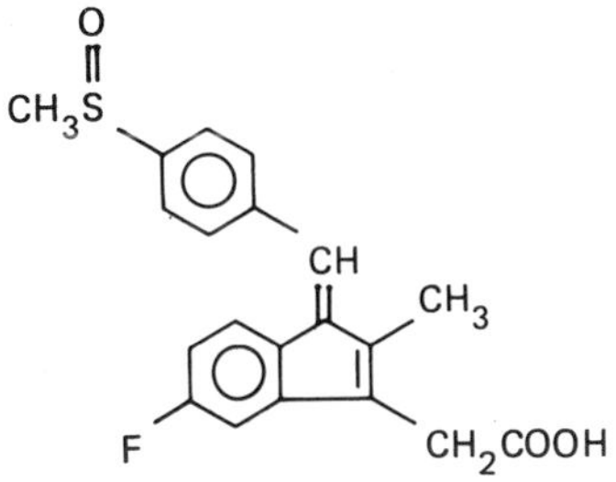
CH_3S
CH
CH_3
F
CH_2COOH

(20)

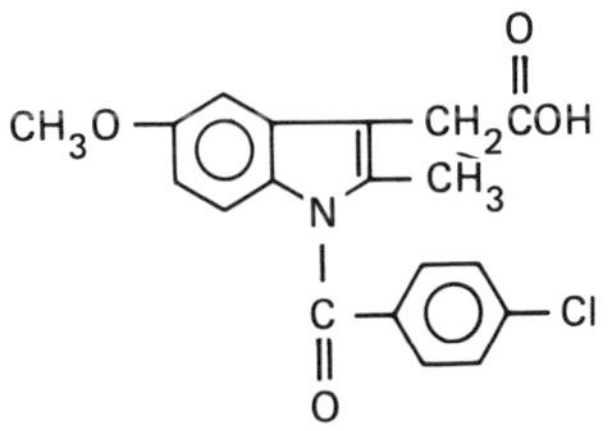
CH_3O
CH_2COH
CH_3
N
C
Cl

(21)

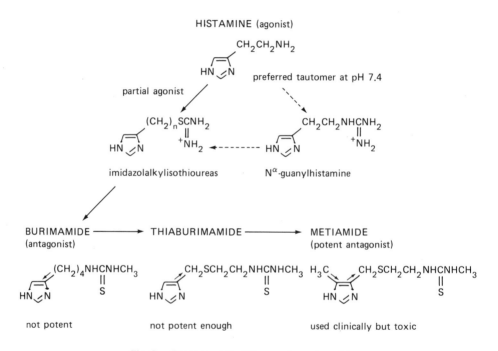

The development of the H$_2$-receptor antagonists has been extensively documented and is outlined in Fig. 5. in some detail because sulphur was introduced

Fig. 5 — Development of H$_2$-receptor antagonists.

specifically to influence electron distribution (Mitchard *et al.*, 1989). It was initially assumed that the structure of the antagonist would be related to that of the agonist histamine. Attempts were therefore made to retain the essential character of the histamine molecule, including the basic nature of the side-chain nitrogen. Partial agonist activity was observed when N^α-guanylhistamine was tested. It was then shown that the imidazole alkylisothioureas, which had previously been synthesized, also showed partial agonist activity. The antagonist and agonist activities were eventually separated when the isothiourea substituent was re-arranged as in burimamide. This re-arrangement fundamenally changes the character of the side-chain nitrogen atoms; they are no longer basic in character. Although a pure antagonist, burimamide was not sufficiently potent for clinical use. It was therefore argued that if a sulphur atom were introduced into the alkyl side-chain this would favour the tautomeric form of the imidazole ring preferred in histamine. The resulting compound, thiaburimamide, was indeed more potent than burimamide but still not potent enough. It was further argued that introduction of an electron-donating methyl group into the 2 position of the imidazole ring would further increase the proportion of the preferred tautomeric form. This produced a compound named metiamide which was sufficiently potent to be used for the treatment of duodenal ulcer. Unfortunately it was too toxic for general clinical use. However, the molecular requirements for H_2-receptor antagonists had been defined and subsequently other molecules were synthesized. Four of these compounds are now in use throughout the world. All contain the thioether alkyl chain but none contains the thiourea substituent and only one, cimetidine, contains the imidazole ring. In two, famotidine and nizatidine, the imidazole has been replaced by a thiazole ring.

All four compounds are oxidatively metabolized to the corresponding inactive sulphoxide compounds. Recent work on H_2-receptor antagonists has demonstrated that there is no fundamental requirement for a sulphur atom in an H_2-receptor antagonist, as excellent activity has been obtained in molecules containing no sulphur.

4.5 CONCLUDING REMARKS

This chapter has attempted to review the many and varied uses of sulphur and its derivatives in industry and medicine. In terms of total sulphur consumption, agriculture and industry account for the majority of the usage. This extensive use of sulphur exposes most biological systems, which includes microorganisms, plants and animals, either to products containing sulphur, for example elemental sulphur, sulphate fertilizers, sulphur pesticides and sulphonated detergents, or to waste products such as sulphur dioxide, bisulphites and sulphonated lignin effluent from pulp mills. In addition, use of sulphur in medicine leads to the intentional exposure of man to this class of chemicals. This requires that we have a clear understanding of the effect and fate of sulphur compounds in various living systems. These aspects of 'sulphur xenobiochemistry' are discussed in detail in various chapters in this and subsequent volumes of this series.

ACKNOWLEDGEMENT

We wish to acknowledge the considerable help given to us by Julie Friend in the preparation of the manuscript.

REFERENCES

Anderson, J. W. (1978). In: *Sulphur in Biology*, The Institute of Biology's Studies in Biology, no. 101, Edward Arnold Ltd., London, pp. 36–48.

Barnes, M. D. (1965). Aspects of sulphur research and potential applications. In B. Meyer (ed.), *Elemental Sulphur: Chemistry and Physics*, Wiley, New York, pp. 357–373.

Biel, J. H., Bopp, B. and Mitchell, B. D. (1978). Chemistry and structure–activity relationships of psychotropic drugs. In W. G. Clark and J. Guidise (eds.), *Principles of Psychopharmacology*, Academic press, New York, pp. 140–168.

Bonner, D. P. and Sykes, R. B. (1984). Structure activity relationshipsa among the monobactams. *J. Antimicrob. Chemotherap.*, **14**, 313–327.

Boswell, C. (1988). The contrast of elemental sulphur fertilizer development in Europe and the Pacific. *Sulphur in Agriculture*, Vol. 12, Published by The Sulphur Institute, pp. 22–23.

Braun, H. (1988). Food and agriculture organisation sulphur network. *Sulphur in Agriculture*, Vol. 12, The Sulphur Institute, pp. 3–5.

Clissold, S. P., Todd, P. A. and Campoli-Richards, D. M. (1987). Imipenem/cilastatin — a review of its antibacterial activity, pharmacokinetic properties and therapeutic efficacy. *Drugs,* **33**, 183–241.

Cooper, R. D. G. and Spry, D. O. (1972). Rearrangements of cephalosporins and penicillins. In E. H. Flynn (ed.), *Cephalosporins and Penicillins — Chemistry and Biology*, Academic Press, New York and London, pp. 183–245.

Damani, L. A. (1987). Metabolism of sulphur-containing drugs. In D. J. Benford, J. W. Bridges and G. G. Gibson (eds.), *Drug Metabolism — from Microbes to Man*, Taylor & Francis, London, pp. 581–603.

Fieser, L. F. and Fieser, M. (1963). In *Topics in Organic Chemistry*, Reinhold, New York, pp. 223–228.

Fike, H. L. (1972). Some potential applications of sulphur. In D. J. Miller and T. K. Wiewiorowski (eds.), *Sulphur Research Trends*, American Chemical Society, Advances in Chemistry Series, No. 110, Washington, DC, pp. 208–224.

Hutson, D. H. and Logan, C. J. (1983). Sulphur in pesticide action. In S. C. Mitchell and R. H. Waring (eds.), *Sulphur in Xenobiotics*, Birmingham University press, Birmingham, pp. 59–66.

Klein, P. (1966). Zur Ideengeschichte der chemotherapeutischen. Fruhperiode. *Dtch. Med. Wochenschr.*, **91**, 2281–2284.

Mitchard, M., McIsaac, R. and Bell, J. A. (1989). H_2-receptor antagonists, their development and comparative clinical pharmacokinetics. In L. A. Damani (ed.), *Sulphur Containing Drugs and Related Organic Compounds: Chemistry, Biochemistry and Toxicology*, Vol. 3, Ellis Horwood, Chichester, in press.

Neu, H. C. (1985). Penicillins. In G. L. Mandell, R. G. Douglas and J. E. Bennett (eds.), *Principles and Practice of Infectious Diseases*, 2nd edn., Wiley, New York, pp. 166–180.

Newall, C. E. (1985). Injectable cephalosporin antibiotics: cephalothin to ceftazidine. In S. M. Roberts and B. J. Price (eds.), *Medicinal Chemistry, The Role of Organic Chemistry in Drug Research*, Academic Press, New York, pp. 209–226.

Nor, Y. M. and Tabatabai, M. A. (1976). Extraction and colorimetric determination of thiosulfate and tetrathionate in soils. *Soil Sci.*, **122**, 171–178.

O'Shaughnessy, M. T. (1958). Xanthation of cellulose. In P. H. Groggin (ed.), *Unit Processes in Organic Synthesis*, McGraw-Hill, New York, pp. 745–746.

Otten, H. (1986). Domagk and the development of the sulphonamides. *J. Antimicrob. Chemother.*, **17**, 689–696.

Porter, M. (1968). The chemistry of the sulphur vulcanization of natural rubber. In A. V. Tobolsky (ed.), *The Chemistry of Sulphides*, Wiley, New York, pp. 165–183.

Reid, E. E. (1963). In *Organic Chemistry of Bivalent Sulphur*, Vol. V, Chemical Publishing Co. Inc., New York, pp. 427–428.

Rosen, J. D., Magee, P. S. and Casida, J. E. (1981). In *Sulphur in Pesticide Action and Metabolism*, American Chemical Symposium Series 158, Washington, D. C. pp. ix–x.

Roth, H. J. and Kleemann, A. (1988). In *Pharmaceutical Chemistry*, Vol. 1, *Drug Synthesis*, Ellis Horwood, Chichester.

Shen, T. Y. and Winter, C. A. (1977). Chemical and biological studies on indomethacin, sulindac and their analogues. *Adv. Drug. Res.*, **12**, 90–245.

The Sulphur Institute (1986–1987). *The Sulphur Institute Annual Report, 1986–1987*, 1725 K. Street N. W., Washington, DC 2006.

Tisdale, S. L., Nelson, W. L. and Beaton, J. D. (1985). In *Soil Fertility and Fertilizers*, 4th edn., Macmillan, New York.

Traxler, R. N. (1961). In *Asphalt, its Compositions, Properties and Uses*, Reinhold, New York, pp. 90–101.

Wainwright, M. (1984). Sulphur oxidation in soils. *Adv. Agron.*, **37**, 349–396.

Worthing, C. R. and Walker, S. B. (1987). In *The Pesticide Manual*, 8th edn., British Crop Protection Council, Thornton Heath.

5

Inorganic sulphur compounds

John Westley
Department of Biochemisty, The University of Chicago, 920 East 58th Street,
Chicago, IL 60637, USA

SUMMARY

1. Available information on the metabolism of inorganic sulphate, sulphite and sulphur dioxide, polythionates, thiosulphate, thiocyanate, elemental sulphur, and sulphide is reviewed.
2. Sulphate, which is excreted as the non-toxic end-product of most sulphur metabolism, is also utilized extensively by activation and transfer in the formation of essential tissue components and the detoxication of various xenobiotics.
3. Sulphite and sulphide are both toxic, the latter at very low concentrations. Both are detoxified by oxidation and to some extent by mechanisms involving sulphur-transferases.
4. Thiosulphate, polythionates and elemental sulphur all contain sulphane sulphur atoms and appear to be active as intermediates in both the oxidation and the formation of sulphide-level sulphur.
5. Thiocyanate, from dietary sources and from cyanide detoxication with sulphane sulphur, has a variety of metabolic effects but is larely excreted unchanged.

5.1 INTRODUCTION

The attempt is here made to provide a compact but reasonably comprehensive account of what is known about the metabolism of inorganic compounds of sulphur in mammalian organisms, with special reference to implications for human biology. Microbial metabolism is addressed only when it seems to have some direct relevance to this central topic, e.g. when possible roles of the enteric bacteria are considered. Fuller views of the extensive field of bacterial inorganic sulphur metabolism, including its relationship to the special metabolism of ruminants, can be found in the book by Roy and Trudinger (1970), the review of Siegel (1975), the monograph by

Anderson (1978), the review by Trudinger and Loughlin (1981) and the recent volume by Huxtable (1986).

The central fact of inorganic sulphur biochemistry is that this element occurs in many oxidation states in animal tissues. These materials range from the most oxidized and least toxic form, sulphate, through sulphite, the polythionates, thiosulphate, thiocyanate, elemental sulphur, and polysulphides to the most reduced and most toxic form, inorganic sulphide. It is perhaps noteworthy that in this listing sulphur often occurs at more than one oxidation level in the same compound. This fact, combined with the ready reactivity of most of the forms (sulphate being the sole major exception), leads frequently to rapid interchange reactions, spontaneous disproportionations and the other complexities that make sulphur biochemistry a special challenge.

Each of the following sections deals with one of the oxidation states of inorganic sulphur that can be found in mammalian tissues, its metabolism, its formation *in vivo* from organic sources, and its (re)incorporation into organic compounds. As appropriate, each section takes up the direct exposure to sulphur inorganics that occurs by virtue of their presence in the air, food and water. A separate chapter on the physiological pool of sulphane sulphur appears in Volume 2, Part B of this series.

5.2 SULPHATE

As implied in the foregoing comments, sulphate is a relatively unreactive form of inorganic sulphur. Quantitatively, however, it is by far the most important form of sulphur in mammalian biology. For example, the participations of sulphate esters in proteoglycan structure, in the sulpholipids of nerve tissue, and in detoxication reactions of many xenobiotics are all processes conducted on a large scale.

The sulphur atom in sulphate is at its highest oxidation level, with a formal valence of +6. Sulphate is therefore not further oxidized *in vivo* and is in fact the principal form excreted. Since mammals also lack the ability to reduce sulphate to any lower oxidation state, sulphate metabolism requires discussion primarily in terms other than those of oxidation and reduction. Apart from a few reactions that produce sulphate *de novo* from other forns, the mammalian biochemistry of sulphate is dominated by activation and transfer reactions.

5.2.1 Activation

Sulphate ion in mammalian tissues is converted by a two-step process to the sulphuryl group of 3'-phosphoadenosine 5'-phosphosulphate (PAPS), which is the sulphuryl donor in sulphate transfer reactions (De Meio, 1975; Segel *et al.*, 1987). The first of the activating reactions is catalysed by ATP-sulphurylase (EC 2.7.7.4):

$$SO_4^{2-} + MgATP \rightleftharpoons APS + MgPPi \tag{1}$$

APS is adenosine 5'-phosphosulphate, the analogue of ADP having the terminal (β) phosphate replaced by sulphate. ATP-sulphurylase has been highly purified from animal, plant, and fungal sources (Burnell and Roy, 1978; Shaw and Anderson, 1972; Tweedie and Segel, 1971). The most extensive studies have been reported for the *Penicillium* enzyme, which catalyses the reaction by a type of mechanism that requires the simultaneous presence of both substrates on the enzyme (Seubert *et al.*,

1980, 1985). Recent studies on the enzyme isolated from rat chondrosarcoma (Geller, Westley and Schwartz, unpublished observations) provide evidence for a somewhat differently detailed mechanism of the same general type, like that previously proposed for the rat liver enzyme (Burnell and Roy, 1978), with ATP as the leading substrate and APS the final product released.

APS

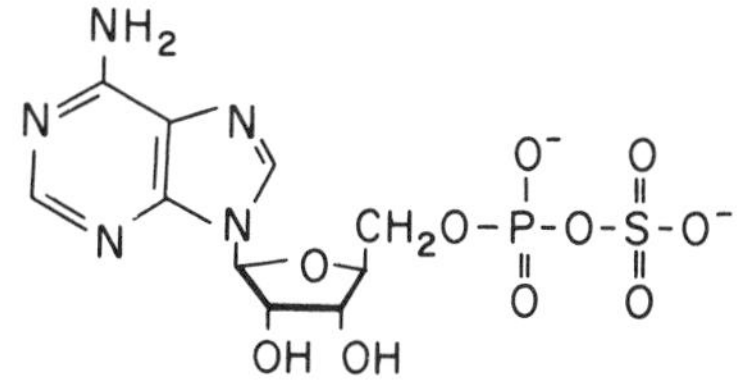

PAPS

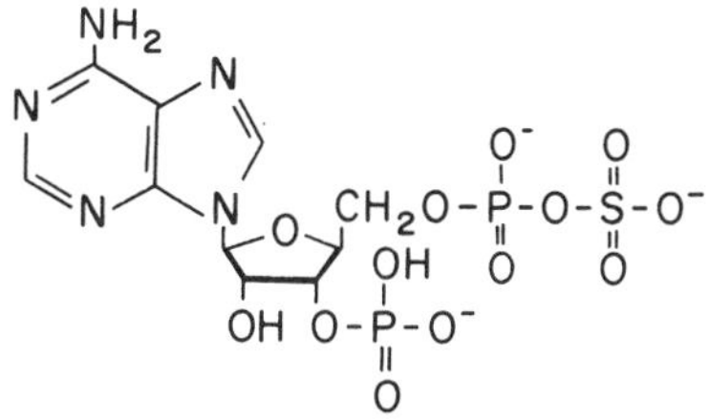

PAP

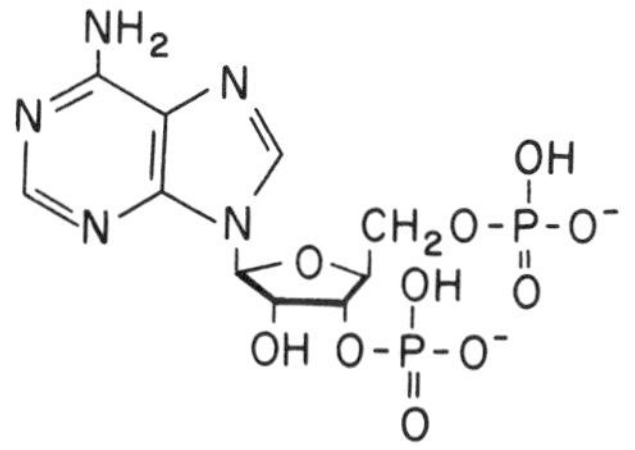

The phosphoryl–sulphuryl anhydride bond in APS is highly reactive, and the equilibrium position of reaction (1) is very far to the left as written. As usual in such circumstances, it is presumed that APS production is made feasible *in vivo* by the action of the ubiquitous inorganic pyrophosphatase on the other product of the reaction:

$$PPi + H_2O \rightleftharpoons 2Pi + 2H^+ \tag{2}$$

In addition, the further reaction of APS, catalysed by APS kinase (EC 2.7.1.25), is also strongly exergonic. Even so, considering the extreme lability of APS, it may well be that it is channelled directly to the kinase reaction without being released into the bulk solvent (Seubert *et al.*, 1985). In this connection, it is noteworthy that ATP-sulphurylase and APS kinase copurify extensively from homogenates of rat chondrosarcoma (Geller *et al.*, 1987), possibly implying the occurrence of sites for catalysis of the two reactions on the same enzyme molecule.

APS kinase catalyses the reaction

$$APS + MgATP \rightleftharpoons PAPS + MgADP \tag{3}$$

PAPS (3'-phosphoadenosine 5'-phosphosulphate) is sufficiently stable that the equilibrium position of reaction (3) is far to the right as written. Thus an overall reaction activating sulphate to PAPS with expense of three phosphoric anhydride bonds is accomplished. Unlike the rat chondrosarcoma enzyme, the kinase as isolated from plant (Burnell and Anderson, 1973) or fungal (Renosto et al., 1984) sources is a protein separate and distinct from ATP-sulphurylase. The formal mechanism, in those cases where it has been investigated (Renosto et al., 1984; (Geller, Westley and Schwartz, unpublished observations) requires the sequential binding of the substrates before release of either product.

There have been recent reports in the neurochemical literature on the development and distributions of the sulphate-activating enzymes in mammalian brain tissue (Brion et al., 1987; Matsuo et al., 1987a,b).

5.2.2 Transfer
Mammalian tissues contain a variety of sulphotransferases (EC 2.8.2.–) which catalyse the use of PAPS as a sulphuryl donor to the hydroxyl or amino groups of a variety of acceptor substrates (De Melo, 1975; Jakoby et al., 1980; Ramaswamy and Jakoby, 1987). The latter include complex lipids as well as such small molecules as phenols and hydroxysteroids and also the glycosaminoglycans in large macromolecules such as chondroitin and heparin. The specificity of the detoxifying sulphotransferases tends to be broad, that of the glycosaminoglycan sulphotransferases and presumably the lipid sulphotransferases much more restricted. In all cases, the reaction catalysed is that shown in eqn. (4), where PAP is the 3'-phosphoric ester of AMP:

$$PAPS + ROH \text{ (or } RNH_2) \rightleftharpoons ROSO_3^- \text{ (or } RNHSO_3^-) + PAP \tag{4}$$

5.2.3 Sulphuric ester hydrolysis
The metabolic degradation of sulphuric esters is catalysed by a group of sulphatases (sulphohydrolases) (EC 3.1.6.–) (Dodgson and Rose, 1975; Roy, 1987). Some of these are specific only for whole classes of compounds (e.g. the arylsulphatases A, B, and C); others are much more highly specific (e.g. the particular glycosaminoglycan oligomer sulphatases).

In all cases, the reaction catalysed is simple hydrolysis according to

$$ROSO_3^- \text{ (or } RNHSO_3^-) + H_2O \rightleftharpoons ROH \text{ (or } RNH_2) + SO_4^{2-} + H^+ \tag{5}$$

5.2.4 Sulphate synthesis
In addition to its hydrolytic release from sulphuric esters and sulphamates, inorganic sulphate is produced in physiological oxidations of inorganic forms at lower oxidation states (see below). It is also formed in the oxidative metabolism of sulphur-containing organics, with the methionyl and cyst(e)inyl residues of proteins as the principal bulk source (Fig. 1). The sulphur of cysteine arising from proteins or glutathione by hydrolysis, or from methionine by transsulphuration, is oxidized to sulphate via cysteine sulphinate, the unstable β-sulphinyl pyruvate, and sulphite. An

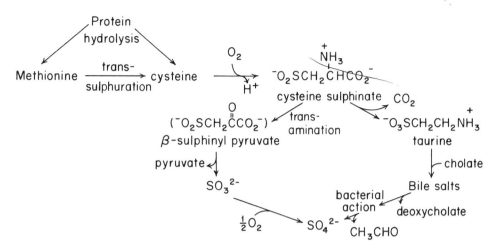

Fig. 1 — Main pathways of sulphur amino acid metabolism to sulphate.

important alternative route of metabolism is via the aminosulphonate taurine, which is conjugated with steroid derivatives to form bile salts. Secreted into the intestine, these are subjected to metabolism by the enteric bacteria, which transaminate the taurine in a process that ultimately yields acetaldehyde and free sulphate (Singer, 1975; Cooper, 1983; Huxtable, 1986).

5.2.5 Transport
Free sulphate anion, sometimes thought not to be well taken up by the intestine, is in fact well enough absorbed to exert a substantial sparing action on the dietary requirement for methionine/cysteine, some of which must otherwise be metabolized to provide sulphate. Sulphate can supply up to 30% of the total requirement for sulphur in rat diets. Entry into peripheral tissue cells is via a carrier (Esko *et al.*, 1986) and there is also energy-dependent active transport of sulphate (and thiosulphate) in the proximal kidney tubule (Ullrich *et al.*, 1980). Moreover, sulphate (and sulphite) can be taken up into liver mitochondria by the dicarboxylate carrier in exchange for phosphate (Crompton *et al.*, 1974; Stipani *et al.*, 1980).

5.3 SULPHITE
As noted in Fig. 1, sulphite is a normal intermediate of amino acid metabolism in mammals. It also may be taken in as such from dietary sources and as respired sulphur dioxide. *In vivo*, as in all neutral buffered aqueous systems, SO_2 reacts very rapidly and completely with water to produce a mixture of HSO_3^- and SO_3^{2-} (Petering, 1977), according to

$$SO_2 + H_2O \rightleftharpoons H^+ + HSO_3^-$$

$$HSO_3^- \overset{pK\ 7.2}{\rightleftharpoons} SO_3^{2-} + H^+$$

(6)

However, in the bloodstream it is found only as the *S*-sulpho derivatives formed by further reaction of sulphite with disulphide bonds according to eqn. (7) (Gunnison and Benton, 1971):

$$RSSR + SO_3^{2-} \rightleftharpoons RS^- + RSSO_3^-$$ (7)

Sulphite, containing sulphur with a formal valence of $+4$, is not subject to reduction to lower inorganic forms in mammalian tissues. Its ultimate disposal is by enzyme-catalysed oxidation to sulphate. Practically all the sulphur retained from sulphur dioxide exposure is excreted as urinary sulphate (Yokoyama *et al.*, 1971).

5.3.1 Oxidation

Sulphite oxidase (EC 1.8.3.1) is a mitochondrial enzyme present at high levels in liver and kidney, with lesser amounts in other tissues (Rajagopalan and Johnson, 1977; Trudinger and Loughlin, 1981). In liver mitochondria, at least, the enzyme is confined to the intermembrane compartment, where its substrate *in vivo* is cytochrome *c*. In intact mitochondria, sulphite oxidation thus causes phosphorylation of ADP with a P:O ratio of 1.0 (Cohen *et al.*, 1973; Oshino and Chance, 1975). For this reason, the sulphite oxidase reaction is an unusual one — an energy-yielding detoxication process. Acute exposure to SO_2 (or injected HSO_3^-/SO_3^{2-}) in large doses can be lethally toxic to sulphite oxidase-depleted rats (Cohen *et al.*, 1973). Total sulphite oxidase deficiency in humans is a lethal condition even without any exogenous exposure (Shih *et al.*, 1977). There is considerable species variation in both sulphite clearance and sulphite oxidase activity measured *in vitro* (Gunnison *et al.*, 1977).

Sulphite oxidase from all sources contains both molybdenum and an iron porphyrin that are essential for activity. Each enzyme protomer contains both cofactors but can be cleaved by tryptic hydrolysis into separate haem- and Mo-containing fragments (Johnson and Rajagopalan, 1977). The available evidence suggests that sulphite *in vivo* reacts first with the bound molybdenum atom, the electrons being then passed on through the cytochrome *b*-like haem prosthetic group to substrate cytochrome *c* and from there through the electron transport system to oxygen.

5.3.2 Sulphite as a nucleophile

Petering's review (1977) serves as a reminder that HSO_3^-/SO_3^{2-} is a reactive and versatile nucleophile. In addition to the sulphitolysis of disulphides and the formation of bisulphite addition products of carbonyl groups, ten documented reactions of sulphite with biomolecules under various conditions approximating the physiological in pH and temperature, if not always in concentration, are listed. Some of these reactions, such as that with the FAD moieties of various flavoenzymes, could have major consequences *in vivo*. Not mentioned, but also of potential significance, is the rapid reaction of sulphite with polythionates and persulphides to form lower polythionates and thiosulphate (see below).

5.3.3 Sulphite *in vivo*

As noted above, the concentrations of free sulphite in the blood or other tissues are ordinarily very low, with practically all the potential sulphite present as *S*-sulphonate compounds. However, in the sulphydryl-rich (largely glutathione) intracellular environment, the equilibrium indicated in eqn. (7) must be displaced toward the left. It is sulphite itself, not the S-sulphonates, that is the substrate of sulphite oxidase. Tissue distributions of S-sulphonates in rabbits exposed to sulphur dioxide at occupational standard levels of 10 ppm showed no substantial transport of exogenous sulphite beyond the respiratory tissues that are the SO_2 absorption sites (Gunnison *et al.*, 1981). However, both the quantity and the quality of the tracheobronchial secretions are much affected by SO_2 exposure. Moreover, both the energy metabolism of lung tissue and the ability to produce inducible detoxifying enzymes for other toxic materials (Husain and Dehnen, 1978) may be severely inhibited. Deleterious effects on tissue ATP levels are negatively correlated with the sulphite oxidase content of the tissue; the energy metabolism of the oxidase-rich hepatocytes is insensitive to sulphite (Beck-Speier *et al.*, 1985).

Sulfite *in vivo* may also be a sulphur-acceptor substrate for 3-mercaptopyruvate sulphurtransferase (EC 2.8.1.2), forming thiosulphate. The same product is formed in the reactions of sulphite with polythionates and persulphides, including the sulphur-substituted rhodanese (thiosulphate : cyanide sulphurtransferase, EC 2.8.1.1). Sulphite thus has access to the physiological sulphane pool, and it is entirely possible that some sulphite is detoxified initially by these means. The central sulphur atom of thiosulphate (the sulphite sulphur atom) is also excreted as urinary sulphate (see below).

5.4 POLYTHIONATES

Although polythionates ($^-O_3SS_xSO_3^-$) are readily formed by either chemical or biological oxidation of thiosulphate (Fig. 2), notably by the *Thiobacilli*, they seem

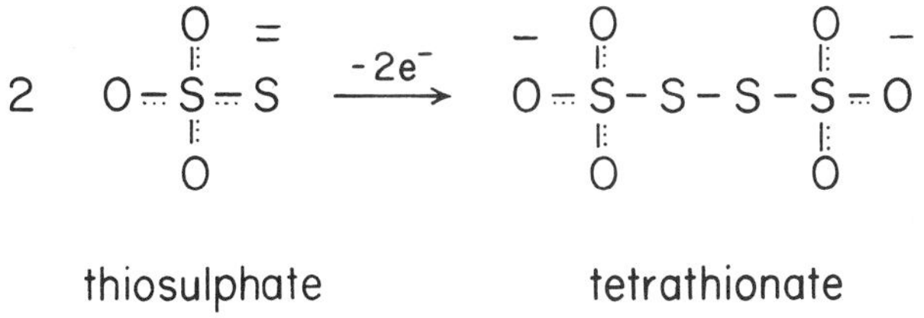

thiosulphate   tetrathionate

Fig. 2 — Oxidation of thiosulphate.

not to occupy a prominent place in mammalian biochemistry. Yet, intraperitoneal injection of SSO_3^{2-} into rats rapidly gives rise to radioactive serum polythionates from tri- through hexathionate ($x = 1/4$ in the above formulation) (Schneider and Westley, 1969), apparently indicating that these are metabolites as normal as thiosulphate itself. In contrast, tetrathionate, in particular, is known to be nephrotoxic in rather large doses (Roy and Trudinger, 1970). This action is related to the ready reactivity of all the polythionates (except trithionate) with good thiophiles.

Fig. 3 illustrates the process with tetrathionate and the thiolate anions of two protein sulphydryl groups. The overall reaction (eqn. (8)) is simply the reduction of tetrathionate to thiosulphate by two thiol groups:

$$S_4O_6^{2-} + 2RS^- \rightarrow 2S_2O_3^{2-} + RSSR \tag{8}$$

It is the loss of sulphydryl groups rather than the production of thiosulphate that results in the toxic effects. Clearly, polythionates are not totally lacking in either biological presence or potential for important effects. Nevertheless, little is known about their participation *in vivo*.

5.4.1 The sulphur atoms of polythionates

It will be noted that the polythionates all contain two highly oxidized sulphur atoms as well as one or more atoms at the sulphane level — divalent sulphur bonded only to other sulphur. The sulphane atoms are increasingly electrophilic as the length of the internal polysulphide chain is increased. Hexathionate, pentathionate and tetrathionate all react rapidly under mild conditions with such good thiophiles as sulphydryl groups (yielding persulphides, RSSH), cyanide anion (yielding thiocyanate, SCN^-), and sulphite (yielding thiosulphate, SSO_3^{2-}). The single sulphane atom in trithionate ($^-O_3SSSO_3^-$) requires somewhat more vigorous conditions for its abstraction by the nucleophiles. Similarly, hexathionate and pentathionate are extremely sensitive to hydroxide ions, decomposing in alkaline solution to deposit elemental sulphur and to generate the next lower polythionate. Tetrathionate and trithionate are progressively more resistant to nucleophilic attack by OH^-, perhaps in part because alkaline decomposition of these materials yields only soluble products (SSO_3^{2-}, SO_3^{2-}, and SO_4^{2-}, depending somewhat on the alkalinity of the solution). Dithionate ($^-O_3SSO_3^-$), which contains no sulphane atom, does not exhibit any of these characteristic reactivities and is not properly considered a polythionate.

5.4.2 Participation *in vivo*

The rather meagre literature on polythionates in mammalian systems is perhaps sufficient to suggest that the topic deserves more attention. These reactive compounds, which are evidently in a rapid equilibrium with thiosulphate and other components of the physiological sulphane pool *in vivo* (see Volume 2, Part B, Chapter 3 of this series), may have been so little studied in biological systems because of their rather elusive chemical nature. Except for tetrathionate, they are not available commercially and much of the conventional methodology for dealing with questions of purity, stability and concentration appears equivocal. Nevertheless, polythionates are not difficult to make (Roy and Trudinger, 1970; Brauer, 1963). Moreover, rapid liquid chromatography on columns of activated carbon (Chapman and Beard, 1973), anion exchangers (Takano *et al.*, 1984) or by reversed-phase ion pairing (Rabin and Stanbury, 1985), or even on microscale thin layer plates in systems that separate thiosulphate, tetrathionate and elemental sulphur (Westley and Westley, 1984), can provide excellent separations. Non-specific detection by short wavelength ultraviolet absorption, conductivity, or Ag^+ reactivity can be used to monitor these separations. Tuovinen and Nickolas (1977) have also reported that a nitrate 'ion-specific' electrode responds better to tetrathionate and trithionate than

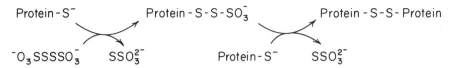

Fig. 3 — Reaction of tetrathionate with protein sulphydryl groups.

to nitrate. Differential pulse polarography provides a more specific detection system useful for analysis of some polythionate mixtures, but direct analysis of solutions containing all the common polythioantes, tri- through hexa-, still poses some problems (Takano *et al.*, 1984).

It thus appears that, although the role(s) of polythionates in mammalian biology have not been much investigated, adequate tools for the conduct of such research may now be available. Like other components of the physiological sulphane pool, polythionates may prove to be involved not only in the detoxication of cyanide (and possibly of sulphite and other nucleophilic xenobiotics as well) but also in the system that provides sulphur for the synthesis of iron–sulphur centres.

5.5 THIOSULPHATE

Like the polythionates, thiosulphate contains both highly oxidized sulphur and sulphane sulphur. Like sulphate anion, thiosulphate anion is a tetrahedral resonance structure having a central (or 'inner') sulphur atom with a formal valence of $+6$. In thiosulphate, however, there is also an 'outer' sulphur atom having a valence of -2, with the three equivalent oxygen atoms in the remaining outer positions (Fig. 4). The

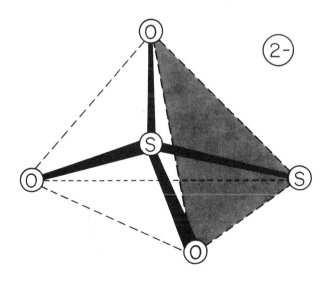

Fig. 4 — Structure of the thiosulphate dianion.

overall net charge of -2 is distributed over all the outer atoms. The sulphur–sulphur bond of thiosulphate is approximately 1.4 times as long as the sulphur–oxygen bonds (Laur, 1972). Chemical cleavage of the S–S bond is commonly by an electrophile-assisted nucleophilic attack (Kice, 1971).

5.5.1 Metabolic fates of thiosulphate sulphur atoms

The inner and outer sulphur atoms of thiosulphate are quite distinct chemically, and either of them may be exclusively labelled with ^{35}S. This fact has greatly aided studies of thiosulphate metabolism. The inner sulphur atom of thiosulphate injected into a rat is rapidly oxidized to sulphate, which is largely excreted as such. The outer sulphur atom, however, is retained for much longer periods and is even incorporated into the cystine of hair during the course of weeks (Skarzynski et al., 1959; Schneider and Westley, 1969). As a result of this incorporation, and doubtless of additinal reactions of this sulphane sulphur with other nucleophiles (e.g. cyanide), it cannot all be accounted for as excreted sulphate, even in the long run.

5.5.2 Thiosulphate-cleaving enzymes

Two sulphurtransferases that cleave thiosulphate are known in mammalian systems: rhodanese and thiosulphate reductase. It appears that one or the other of these enzymes must be involved in the intial step of all metabolic disposition of thiosulphate.

5.5.2.1 Rhodanese

Rhodanese (thiosulphate : cyanide sulphurtransferase, EC 2.8.1.1), originally discovered as the result of a search for a cyanide-detoxifying catalyst, has been much studied for its phylogenetic, tissue, and subcellular distributions (Koj, 1968, 1980) as well as its chemical mechanism (Schlesinger and Westley, 1974; Westley, 1977) and detailed structure (Russell et al., 1978; Ploegman et al., 1978). Research on this enzyme has been reviewed in various contexts (Sörbo, 1975; Volini and Alexander, 1981; Westley, 1973, 1980, 1981a; Hol et al., 1983; Westley et al., 1983; Finazzi-Agro et al., 1971; Cerletti, 1986).

Rhodanese is a soluble enzyme of the mitochondrial matrix in mammalian tissues. It is particularly abundant in liver and kidney but is also present in other organs. As a protein, it is a double-domain structure of 33 000 daltons, with the catalytically active site located in a cleft between the domains.

Thiosulphate anion (or other anion with a terminal sulphane sulphur) forms a complex with positively charged residues in the active site. The cationic site also serves as an electrophilic group, polarizing the sulphur–sulphur bond to be cleaved. Bond scission is accomplished in this complex by nucleophilic attack of an enzymic sulphydryl group to form a persulphide, which is a covalently substituted enzyme intermediate. Sulphite (or the analogous product) is then discharged from the enzyme.

The sulphur-substituted enzyme, which is the form isolated by the common purification procedures (Horowitz and DeToma, 1970; Horowitz, 1978; Westley, 1981b), reacts with a thiophilic acceptor substrate to produce the final product and to regenerate the free enzyme. When cyanide anion is the acceptor substrate, the final product is thiocyanate. Monothiols (RSH), which are rather poor acceptor substrates, produce persulphides (RSSH). Dithiols such as lipoate are much better acceptors, yielding transient persulphides which undergo intramolecular reaction as shown in Fig. 5 to produce the corresponding disulphides and hydrosulphide anion as

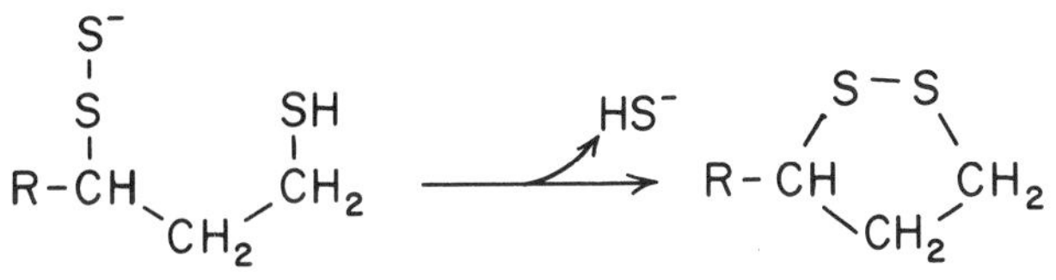

Fig. 5 — Intramolecular reaction releasing sulphide from the persulphide of a dithiol such as lipoate.

final products. Rhodanese can thus serve as a dithiol-based thiosulphate reductase. However, there is evidence indicating that two-thirds of the thiosulphate reduction in liver is catalysed by a monothiol-dependent thiosulphate reductase rather than by rhodanese with a dithiol acceptor substrate (Sörbo, 1964; Koj, 1968).

5.5.2.2   Thiosulphate reductase

Thiosulphate reductase (thiosulphate:thiol sulphurtransferase, EC 2.8.1.3) is ubiquitous in mammalian tissues (Koj *et al.*, 1977) and is not so strictly confined to the mitochondria as is rhodanese (Koj *et al.*, 1975). It has, however, so labile an activity that it has not been purified substantially from any mammalian source. All the mechanism work has been carried out with the yeast enzyme, which can be purified to apparent homogeneity (Chauncey and Westley, 1983a; Chauncey *et al.*, 1987).

The isolated yeast enzyme differs markedly from mammalian rhodanese in molecular properties, being half the size and possessing no essential sulphydryl group. Nevertheless, thiosulphate reductase, like rhodanese, catalyses the cleavage of inorganic thiosulphate or an organic thiosulphonate anion ($RS(O)_2S^-$) with transfer of the sulphane atom to a thiophilic acceptor substrate. However, there is a difference in acceptor specificity. Thiosulphate reductase does not transfer directly to cyanide, but requires a thiol as acceptor (Koj, 1968; Uhteg and Westley, 1979).

The chemical mechanism of action appears to be very similar to that of rhodanese even though the kinetic mechanisms differ greatly (Chauncey and Westley, 1983b). Yeast thiosulphate reductase, lacking the active site sulphydryl group of rhodanese, utilizes the sulphydryl group of the acceptor substrate as a cleaving nucleophile. There is thus no covalently substituted enzyme intermediate for thiosulphate reductase, and the cleavage requires the prior presence of the acceptor substrate on the enzyme. The immediate product is the persulphide of the acceptor (eqn. (9)). Inorganic hydrosulphide is then produced in an uncatalysed reaction of the persulphide with excess thiol substrate (eqn. (10)).

$$SSO_3^{2-} + RSH \underset{\longleftarrow}{\overset{\text{enzymic}}{\rightleftharpoons}} RSSH + SO_3^{2-} \tag{9}$$

$$\text{RSSH} + \text{RS}^- \underset{}{\overset{\text{uncatalysed}}{\rightleftharpoons}} \text{HS}^- + \text{RSSR} \tag{10}$$

It is not yet known whether the mammalian thiosulphate reductase functions by the same mechanism as the yeast enzyme.

5.5.3 Metabolic sources of thiosulphate

Thiosulphate is a physiological material that occurs in normal human urine (Sörbo *et al.*, 1980), despite some literature comments to the contrary. Practically all thiosulphate *in vivo* must be, at least in part, a product of cysteine metabolism. There is rather little dietary thiosulphate and, in any case, thiosulphate decomposes rapidly to elemental sulphur and sulphurous acid at pH values of the gastric contents.

5.5.3.1 The sulphonyl atom and the sulphane atom

The largest-scale pathway of cysteine metabolism is via cysteine sulphinate and β-sulphinyl pyruvate to sulphite and pyruvate (see Fig. 1). If an appropriate source of sulphane sulphur is available, the sulphite so derived can react with it to produce thiosulphate.

The sulphur-substituted rhodanese is one form of sulpane sulphur that is very reactive with sulphite (Sörbo, 1957a). Another source is 3-mercaptopyruvate (see below), which is obtained from cysteine by transamination. A third potential source is the cysteine persulphide formed by the asymmetric cleavage of cystine by γ-cystathionase (cystathionine γ-lyase, EC 4.4.1.1), but the intracellular concentration of free cystine is so low that the actual contribution from this precursor is probably negligible.

The principal *de novo* source of the sulphane sulphur atom of thiosulphate *in vivo* is thought to be the reactions catalysed by 3-mercaptopyruvate sulphurtransferase (EC 2.8.1.2). This enzyme, which has broad phylogenetic and tissue distributions, is bimodally distributed within the cell, with maxima occurring in the mitochondrial matrix and cytosolic fractions (Koj, 1980). Although thiosulphate does not have free diffusion access across the mitochondrial membrane, it is transported by the dicarboxylate carrier (Stipani *et al.*, 1980), as noted in a foregoing section.

The mercaptopyruvate sulphurtransferase isolated from kidney tissue, where it occurs almost entirely within the mitochondrial matrix, has been subjected to bisubstrate kinetic analysis (Jarabak and Westley, 1978, 1980). The mechanism of action requires the simultaneous presence on the enzyme of both the donor and the acceptor substrates but is complicated by an apparent loose specificity for the latter. It was Sörbo (1957b) who showed that sulphite can serve as a sulphur-acceptor substrate for this enzyme, with production of thiosulphate. Moreover, with a thiol as acceptor (even with mercaptopyruvate itself as that thiol) the product is a persulphide which can then react with sulphite to produce thiosulphate:

$$\text{RSSH} + \text{SO}_3^{2-} \rightleftharpoons \text{S}_2\text{O}_3^{2-} + \text{RSH} \tag{11}$$

Furthermore, as noted in early work with crude preparations of this enzyme (Meister *et al.*, 1954; Hylin and Wood, 1959), elemental sulphur and/or heterogeneous protein polysulphides may be produced as well, and these can also react with sulphite. Such observations have been confirmed in studies with the highly purified kidney enzyme cited above.

5.6   THIOCYANATE

Although inorganic thiocyanate is a normal constituent of saliva, blood, and urine, and despite the fact that its administration elicits both a striking hypotensive response and a goitrogenic effect, this ion seems not to have a very active metabolism. Thiocyanate, whether injected (Wood *et al.*, 1947), given orally (Okoh and Pitt, 1982), or subjected to prolonged incubation *in vitro* with oral enzymes and microflora (Tonzetich and Catherall, 1976), is recoverable almost entirely as unchanged thiocyanate. The literature on thiocyanate metabolism prior to 1975 is accessible through Wood's extensive review (1975).

5.6.1   Thiocyanate as a pseudohalide

As Wood (1975) has suggested, much of the physiological handling of thiocyanate appears to reflect its chemical status as a pseudohalide. Thus, thiocyanate competes with iodide in the thyroid, with resultant increased frequency of goitre in patients taking thiocyanate as a hypotensive agent. The same result occurs in populations with chronically high thiocyanate levels caused by the ingestion of foodstuffs (e.g. cassava) containing a high content of cyanogenic glycosides. Cyanide is detoxified *in vivo* primarily by reaction with sulphane sulphur to form thiocyanate.

The excretion of thiocyanate is almost entirely via the kidneys but, despite ready filtration through the glomeruli, efficient reabsorption (possibly as a halide) results in substantial retention and recycling. Again, perhaps predictably, recycled thiocyanate competes with chloride in the secretion of gastric HCl, and thiocyanate tends to accumulate in the gastric contents. It is, however, reabsorbed in the intestine rather than being discharged in the faeces. For example, after oral administration of small doses of ^{14}C-labelled thiocyanate to rats, Okoh and Pitt (1982) found that 90% of the dose could be recovered in the urine during the next five days, almost all of it as unchanged thiocyanate.

5.6.2   Peroxidation of thiocyanate

Despite the indications of metabolic inertness noted above, a small fraction of administered thiocyanate (typically <5%) is oxidized *in vivo* to sulphate and cyanide. The reaction is catalysed by peroxidases, according to

$$SCN^- + 3H_2O_2 \rightarrow SO_4^{2-} + CN^- + 2H_2O \tag{12}$$

It may be noted that this reaction, to the extent that it occurs, can be regarded as a kind of 'slippage' in the cyanide detoxication mechanism, regenerating some of the poison from its detoxication product.

5.6.3   Sources of thiocyanate

The largest source of thiocyanate is the diet (see Chapter 2 in Part A of this volume and Chapter 10 in Part B). Cruciferous vegetables (cabbage, broccoli, kale etc.) are particularly noteworthy for their content of the glucosinolates that produce thiocyanate on hydrolysis by endogenous enzymes. The only important source of *de novo* thiocyanate is the cyanide-detoxifying reaction previously referred to. Small amounts of cyanide are common in most diets because many ordinary foodstuffs contain small quantities of cyanogenic glycosides from which cyanide is released in much the same way that thiocyanates arise from glucosinolates in prepared dishes of cruciferous vegetables. Unfortunately, some plant materials (principally cassava but also sorghums, millets, lima beans etc.) that are important staples in both human food and livestock feed in many tropical cultures contain such high concentrations of cyanogenic glycosides as to constitute a serious, large-scale medical problem (Nartey, 1980) despite both a general awareness of the toxicity and various processing strategems to mitigate it (Dufour, 1985). Moreover, there is HCN in tabacco smoke, and smokers consequently have higher plasma thiocyanate levels than non-smokers. Finally, small amounts of cyanide are produced endogenously by the action of the myeloperoxidase system in phagocytosis by polymorphonuclear leucocytes (Stelmaszyńska and Zgliczynski, 1980). Cyanide from all these sources is detoxified by reaction with sulphane sulphur to form thiocyanate (Westley, 1981a) (also see chapter by Westley in Volume 2, Part B of this series).

5.7 ELEMENTAL SULPHUR

Elemental sulphur, with all its atoms at the sulphane level, is not widely known as a material active in metabolisn. The once popular use of colloidal sulphur injections to treat rheumatoid arthritis was discontinued nearly 50 years ago after a careful review of the available evidence (Comroe, 1939) failed to establish its effectveness. Orally administered colloidal sulphur, but not powdered sulphur, is promptly absorbed, oxidized and almost quantitatively eliminated as urinary sulphate (Greengard and Wooley, 1940). The current medical use of the free element appears to be confined to salves and ointments for topical application to the skin (see chapter 1 by Burt in Volume 3, Part A of this series).

Nevertheless, there are some indications that elemental sulphur may occur physiologically and, furthermore, that such sulphur may play an important metabolic role. The physiological pool of sulphane sulphur that can be labelled readily by injection of rats with SSO_3^{2-} was originally reported to include a substantial portion of the label that behaved as if it were 'protein-coated elemental sulphur'. In fact, an injected soluble complex formed *in vitro* from radioactive elemental sulphur and bovine serum albumin was found to be slowly metabolized to provide sulphur that was incorporated into hair cystine over a long period of time (Schneider and Westley, 1969).

It now appears that the 'protein-coated elemental sulphur' that is formed *in vivo* is a specific complex of serum albumin with some form of the free element (Westley *et al.*, 1983). Such a finding is in accord with Sörbo's (1955) early observations of the catalytic activity of serum albumin for the cyanolysis of colloidal sulphur. It has recently been possible to explore the mechanism of this catalysis (Jarabak and Westley, 1986), and the mode of sulphur binding is under current study. The data thus far support the hypothesis that elemental sulphur bound at specific sites on serum albumin is the main *in vivo* transport form of the sulphane sulphur active in the synthesis of iron–sulphur centres and in cyanide detoxication.

5.8 SULPHIDE

Whereas the oxidized forms of sulphur, like the elemental form, are either non-toxic or toxic only in relatively large doses, inorganic sulphide is a poison as potent as cyanide (Evans, 1967). Beauchamp *et al.* (1984) have provided an exhaustive review of the sulphide toxicology literature. Because of the extreme toxicity, the primary problems of inorganic sulphide metabolism tend to be dominated by the need for avoidance or detoxication. Avoidance in this context involves both the prevention of accidental exposure to H_2S and the prevention of sulphide production by the endogenous reduction of metabolic sulphane sulphur. Detoxication has to do with the metabolic disposal of preformed sulphide from all sources. Moreover, avoidance must include the absence of xenobiotics that yield sulphide *in vivo*. For example, carbonyl sulphide, a gaseous industrial by-product (Morris *et al.*, 1979), as well as a metabolic product of carbon disulphide (Dalvi and Neal, 1978), is toxic primarily because it in turn is metabolized to produce H_2S (Chengelis and Neal, 1979, 1980).

5.8.1 Detoxication

Total avoidance of exposure to sulphide is not a practical goal. This is not simply a matter of industrial contamination of the atmosphere with H_2S. There is a massive microbiological sulphur cycle that annually releases more than 10^8 tons of sulphur into the atmosphere as H_2S (Kellog *et al.*, 1972; Anderson, 1978). Moreover, the anaerobic microflora of the intestinal tract produce substantial amounts of H_2S. Even within our own tissues, the potential for sulphide release occurs in all reactions that can produce persulphides.

Detoxication is accomplished by the action of thiol *S*-methyltransferase as well as by both enzymic and non-enzymic oxidation processes (Weisiger and Jakoby, 1979). Methanethiol, the product of the *S*-methyltransferase reaction with sulphide, is still about one-tenth as toxic as H_2S. However, it too is a substrate for the same enzyme, with ultimate production of the much less toxic thioether dimethyl sulphide, H_3CSCH_3. The enzyme, which has a membrane-associated activity, occurs at high specific activity in the caecal and colonic mucosa as well as the liver, lung and kidney — all locations where it could be expected to exert an important protective effect against sulphide (Weisiger and Jakoby, 1980). Oxidative sulphide metabolism, which is very active in liver mitochondria (Bartholomew *et al.*, 1980), is thought to proceed via polysulphides, thiosulphate, and sulphite to sulphate. Obviously, such oxidation also has a detoxifying effect, and most sulphide sulphur can be recovered as urinary sulphate. In a different approach to limiting sulphide-induced damage, it is interesting that injections of haem, in a dosage that has no observable effect in controls, can completely reverse the effects of prior sulphide injection on mitochondrial porphyrin metabolism in rats (Savolainen *et al.*, 1985).

5.8.2 Avoidance

Apart from the obvious behavioural aspects of sulphide avoidance, there is also the matter of preventing its endogenous release (eqn. (10)) from persulphides generated by such enzymes as thiosulphate reductase, mercaptopyruvate sulphurtransferase, and γ-cystathionase. Sulphide is extremely neurotoxic. Like cyanide, it inactivates the terminal oxidase of the respiratory electon transport system. Even small amounts also cause delayed effects on brain protein synthesis (Elovarra *et al.*, 1978). For this reason, prophylactic rather than merely therapeutic protection would seem to be in order. A role for rhodanese in such prophylactic detoxication has been proposed (Szczepkowski, 1961; Koj and Frendo, 1962; Szczepkowski and Wood, 1967). Rhodanese catalyses the formation of stable, non-toxic thiosulphate from persulphides and sulphite:

$$RSSH + SO_3^{2-} \overset{\text{rhodanese}}{\rightleftharpoons} SSO_3^{2-} + RSH \qquad (13)$$

Although this proposal has been criticized on grounds that rhodanese is confined to the mitochondrial matrix while the persulphide-generating enzymes are not, the fact remains that the mitochondria are also the location of the cytochrome oxidase that is the target to be protected.

ACKNOWLEDGEMENTS

Current research in this laboratory on these topics is supported by research grants GM 30971 from the US National Institutes of Health and DBM 85-12836 from the US National Science Foundation.

REFERENCES

Anderson, J. W. (1978). In *Sulphur in Biology*, University Park Press, Balitmore, pp. 11–22.

Bartholomew, T. C., Powell, G. M., Dodgson, K. S. and Curtis, C. G. (1980). Oxidation of sodium sulfide by rat liver, lungs and kidney. *Biochem. Pharmacol.*, **29**, 2431–2437.

Beauchamp, R. O., Bus, J. S., Popp, J. A., Boreiko, C. J. and Andjelkovich, D. A. (1984). A critical review of literature on hydrogen sulfide toxicity. *CRC Crit. Rev. Toxicol.*, **13**, 25–97.

Beck-Speier, I., Hinze, H. and Holzer, H. (1985). Effect of sulfite on the energy metabolism of mammalian tissues in correlation to sulfite oxidase activity. *Biochim. Biophys. Acta*, **841**, 81–89.

Brauer, G. (1963). In *Handbook of Preparative Inorganic Chemistry*, 2nd edn., Academic Press, New York, pp. 398–403.

Brion, F., Schwartz, J.-C. and Vargas, F. (1987). Properties and localization of the sulfate-activating system in rat brain. *J. Neurochem.*, **48**, 1171–1177.

Burnell, J. N. and Anderson, J. W. (1973). Adenosine 5′-sulphatophosphate kinase activity in spinach leaf tissue. *Biochem. J.*, **134**, 565–579.

Burnell, J. N. and Roy, A. B. (1978). Purification and properties of the ATP sulfurylase of rat liver. *Biochim. Biophys. Acta*, **527**, 239–248.

Cerletti, P. (1986). Seeking a better job for an under-employed enzyme: rhodanese. *TIBS*, **11**, 369–372.

Chapman, J. N. and Beard, H. R. (1973). Fast analytical procedure for the separation and determination of the polythionates found in Wackenroder's solution by high speed liquid chromatography. *Anal. Chem.*, **45**, 2268–2270.

Chauncey, T. R. and Westley, J. (1983a). Improved purification and sulfhydryl analysis of thiosulphate reductase. *Biochim. Biophys. Acta*, **744**, 304–311.

Chauncey, T. R. and Westley, J. (1983b). The catalytic mechanism of yeast thiosulfate reductase. *J. Biol. Chem.*, **258**, 15037–15045.

Chauncey, T. R., Uhteg, L. C. and Westley, J. (1987). Thiosulphate reductase. *Methods Enzymol.*, **143**, 350–354.

Chengelis, C. P. and Neal, R. A. (1979). Hepatic carbonyl sulfide metabolism. *Biochem. Biophys. Res. Commun.*, **90**, 993–999.

Chengelis, C. P. and Neal, R. A. (1980). Studies of carbonyl sulfide toxicity: metabolism by carbonic anhydrase. *Toxicol. Appl. Pharmacol.*, **55**, 198–202.

Cohen, H. J., Drew, R. T., Johnston, J. L. and Rajagopalan, K. V. (1973). Molecular basis of the biological function of molybdenum. *Proc. Natl. Acad. Sci. USA*, **70**, 3655–3659.

Comroe, B. I. (1939). Sulphur therapy in arthritis. *Medicine*, **18**, 203–219.

Cooper, A. J. L. (1983). Biochemistry of sulfur-containing amino acids. *Ann. Rev. Biochem.*, **52**, 187–222.

Crompton, M., Palmieri, F., Capano, M. and Quagliariello, E. (1974). The transport of sulphate and sulphite in rat liver mitochondria. *Biochem. J.*, **142**, 127–137.

Dalvi, R. R. and Neal, R. A. (1978). Metabolism *in vivo* of carbon disulfide to carbonyl sulfide and carbon dioxide in the rat. *Biochem. Pharmacol.*, **27**, 1608–1609.

De Meio, R. H. (1975). Sulfate activation and transfer. In D. M. Greenberg (ed.), *Metabolism of Sulfur Compounds*, Metabolic Pathways, Vol. VII, 3rd edn., Academic Press, New York, pp. 287–358.

Dodgson, K. S. and Rose, F. A. (1975). Sulfohydrolases. In D. M. Greenberg (ed.), *Metabolism of Sulfur Compounds*, Metabolic Pathways, Vol. VII, 3rd edn., Academic Press, New York, pp. 359–431.

Dufour, D. L. (1985). Manioc as a dietary staple: implications for the budgeting of time and energy in the northwest Amazon. In D. J. Cattle and K. H. Schwerin (eds.), *Food Energy in Tropical Ecosystems*, Gordon and Breach, New York, pp. 1–20.

Elovarra, E., Tossavainen, A. and Savolainen, H. (1978). Effects of subclinical hydrogen sulfide intoxication on mouse brain protein metabolism. *Exp. Neurol.*, **62**, 93–98.

Esko, J. D., Elgavish, A., Prasthofer, T., Taylor, W. H. and Weinke, J. L. (1986). Sulfate transport-deficient mutants of Chinese hamster ovary cells. *J. Biol. Chem.*, **261**, 15725–15733.

Evans, C. L. (1967). The toxicity of hydrogen sulfide and other sulfides. *Q. J. Exp. Physiol. Cogn. Med. Sci.*, **52**, 231–248.

Finazzi-Agro, A., Cannella, C., Graziani, M. T. and Cavallini, D. (1971). A possible role for rhodanese. *FEBS Lett.*, **16**, 172–174.

Geller, D. H., Henry, J. G., Belch, J. and Schwartz, N. B. (1987). Copurification and characterization of ATP-sulfurylase and adenosine-5'-phosphosulfate kinase from rat chondrosarcoma. *J. Biol. Chem.*, **262**, 7374–7382.

Greengard, H. and Woolley, J. R. (1940). Studies on colloidal sulfur–polysulfide mixture. *J. Biol. Chem.*, **132**, 83–89.

Gunnison, A. F. and Benton, A. W. (1971). Sulfur dioxide: sulfite. *Arch. Environ. Health*, **22**, 381–388.

Gunnison, A. F., Bresnahan, C. A. and Palmes, E. D. (1977). Comparative sulfite metabolisn in the rat, rabbit, and rhesus monkey. *Toxicol. Appl. Pharmacol.*, **42**, 99–109.

Gunnison, A. F. Zaccardi, J., Dulak, L. and Chiang, G. (1981). Tissue distribution of S-sulfonate metabolites following exposure to sulfur dioxide, *Environ. Res.*, **24**, 432–443.

Hol, W. G. J., Lijk, L. J. and Kalk, K. H. (1983). The high resolution three-dimensional structure of bovine liver rhodanese. *Fundam. Appl. Toxicol.*, **3**, 370–376.

Horowitz, P. M. (1978). Purification of thiosulfate sulfurtransferase by selective immobilization on blue agarose. *Anal. Biochem.*, **86**, 751–753.

Horowitz, P. and DeToma, F. (1970). Improved preparation of bovine liver rhodanese. *J. Biol. Chem.*, **245**, 984–985.

Husain, M. M. and Dehnen, W. (1978). Effect of NO_2 and SO_2 inhalation on benzo(a)pyrene metabolism in rat lung. *Arch. Toxicol.*, **40**, 207–210.

Huxtable, R. J. (1986). In *Biochemistry of Sulfur*, Plenum, New York, pp. 11–61.

Hylin, J. W. and Wood, J. L. (1959). Enzymatic formation of polysulfides from mercaptopyruvate. *J. Biol. Chem.*, **234**, 2141–2144.

Jakoby, W. B., Sekura, R. D., Lyon, E. S., Marcus, C. J. and Wang, J. L. (1980). Sulfotransferases. In W. B. Jakoby (ed.), *Enzymatic Basis of Detoxication*, Vol. II, Academic Press, New York, pp. 199–228.

Jarabak, R. and Westley, J. (1978). Steady-state kinetics of 3-mercaptopyruvate sulfurtransferase from bovine kidney. *Arch. Biochem. Biophys.*, **185**, 458–465.

Jarabak, R. and Westley, J. (1980). 3-Mercaptopyruvate sulfurtransferase: rapid equilibrium-ordered mechanism with cyanide as acceptor substrate. *Biochemistry*, **19**, 900–904.

Jarabak, R. and Westley, J. (1986). Serum albumin and cyanide detoxication. *J. Biol. Chem.*, **261**, 10793–10796.

Johnson, J. L. and Rajagopalan, K. V. (1977). Tryptic cleavage of rat liver sulfite oxidase. *J. Biol. Chem.*, **252**, 2017–2025.

Kellog, W. W., Cadle, R. D., Allen, E. R., Lazrus, A. L., and Martell, E. A. (1972). The sulphur cycle. *Science*, **175,** 587–596.

Kice, J. L. (1971). The sulfur–sulfur bond. In A. Senning (ed.), *Sulfur in Organic and Inorganic Chemistry*, Vol. 1, Dekker, New York, pp. 153–207.

Koj, A. (1968). Enzymic reduction of thiosulfate in preparations from beef liver. *Acta Biochim. Polon.*, **15**, 161–169.

Koj, A. (1980). Subcellular compartmentation and biological functions of mercapto-pyruvate sulphurtransferase and rhodanese. In D. Cavallini, G. E. Gaull and V. Zappia (eds.), *Natural Sulfur Compounds*, Plenum, New York, pp. 493–503.

Koj, A. and Frendo, J. (1962). The activity of cysteine desulfhydrase and rhodanese in animal tissues. *Acta Biochim. Polon.*, **9**, 373–379.

Koj, A., Frendo, J. and Woitczak, L. (1975). Subcellular distribution and intramito-chondrial localization of three sulfurtransferases in rat liver. *FEBS Lett.*, **57**, 42–46.

Koj, A., Michalik, M. and Kasperczyk, H. (1977). Mitochondrial and cytosolic activities of three sulfurtransferases in some rat tissues and Morris hepatomas. *Bull. Acad. Pol. Sci.*, *Ser. Sci. Biol.*, **25**, 1–6.

Laur, P. H. (1972). Steric aspects of sulfur chemistry. In A. Senning (ed.), *Sulfur in Organic and Inorganic Chemistry*, Vol. 3, Dekker, New York, pp. 91–274.

Matsuo, K., Moss, L. and Hommes, F. (1987a). Properties of the 3′-phospoadenosine-5′-phosphosulfate (PAPS) synthesizing systems of brain and liver. *Neurochem. Res.*, **12**, 345–349.

Matsuo, K., Moss, L. and Hommes, F. (1987b). Development of adenosine 5′-triphosphate sulfurylase and adenosine phosphosulfate kinase in rat cerebrum and liver. *Dev. Neurosci.*, **9**, 128–132.

Meister, A., Fraser, P. E. and Tice, S. V. (1954). Enzymatic desulfuration of β-mercaptopyruvate to pyruvate. *J. Biol. Chem.*, **206**, 561–575.

Morris, S. C., Moskowitz, P. D., Sevain, W., Silberstein, S. and Hamilton, L. (1979). Coal conversion technologies: some health and environmental effects. *Science*, **206**, 654–662.

Nartey, F. (1980). Cyanogenesis in tropical foods. In B. Vennesland, E. E. Conn, C. J. Knowles, J. Westley and F. Wissing (eds.), *Cyanide in Biology*, Academic Press, London, pp. 115–132.

Okoh, P. N. and Pitt, G. A. J. (1982). The metabolism of cyanide and the gastrointestinal circulation of the resulting thiocyanate under conditions of chronic cyanide intake in the rat. *Can. J. Physiol. Pharmacol.*, **60**, 381–386.

Oshino, N. and Chance, B. (1975). The properties of sulfite oxidation in perfused rat liver; interaction of sulfite oxidase with the mitochondrial respiratory chain. *Arch. Biochem. Biophys.*, **170**, 514–528.

Petering, D. H. (1977). Sulfur dioxide: a view of its reactions with biomolecules. In S. D. Lee (ed.), *Biochemical Effects of Environmental Pollutants*, Ann Arbor Science, Ann Arbor, MI, pp. 293–306.

Ploegman, J. H., Drent, G., Kalk, K. H. and Hol, W. G. J. (1978). Structure of bovine liver rhodanese. *J. Mol. Biol.*, **123**, 557–594.

Rabin, S. B. and Stanbury, D. M. (1985). Separation and determination of polythionates by ion-pair chromatography. *Anal. Chem.*, **57**, 1130–1132.

Rajagopalan, K. V. and Johnson, J. L. (1977). Biological origin and metabolism of SO_2. In S. D. Lee (ed.), *Biochemical Effects of Environmental Pollutants*, Ann Arbor Science, Ann Arbor, MI, pp. 307–314.

Ramaswamy, S. G. and Jakoby, W. B. (1987). Sulfotransferase assays. In W. B. Jakoby and O. W. Griffith (eds.), *Sulfur and Sulfur Amino Acids. Methods in Enzymology*, Vol. 143, Academic Press, New York, pp. 201–207.

Renosto, F., Seubert, P. A. and Segel, I. H. (1984). Adenosine 5'-phosphosulfate kinase from *Penicillium chrysogenum*. *J. Biol. Chem.*, **259**, 2113–2123.

Roy, A. B. (1987). Arylsulfatases. In W. B. Jakoby and O. W. Griffith (eds.), *Sulfur and Sulfur Amino Acids. Methods in Enzymology*, Vol. 143, Academic Press, New York, pp. 207–217.

Roy, A. B. and Trudinger, P. A. (1970). In *The Biochemistry of Inorganic Compounds of Sulfur*, Cambridge University Press, London, pp. 207–288.

Russell, J., Weng, L., Keim, P. S. and Heinrikson, R. L. (1978). The covalent structure of bovine liver rhodanese. *J. Biol. Chem.*, **253**, 8102–8108.

Savolainen, H., Jappinen, P. and Tenhunen, R. (1985). Reversal of sulfide-induced efects on porphyrin metabolism by endogenous heme. *Res. Commun. Chem. Pathol. Pharmacol.*, **50**, 245–250.

Schlesinger, P. and Westley, J. (1974). An expanded mechanism for rhodanese catalysis. *J. Biol. Chem.*, **249**, 780–788.

Schneider, J. F. and Westley, J. (1969). Metabolic interrelations of sulfur in proteins, thiosulfate and cystine. *J. Biol. Chem.*, **244**, 5735–5744.

Segel, I. H., Renosto, R. and Seubert, P. A. (1987). Sulfate-activating enzymes. In W. B. Jakoby and O. W. Griffith (eds.), *Sulfur and Sulfur Amnino Acids. Methods in Enzymology*, Vol. 143, Academic Press, New York, pp. 334–349.

Seubert, P. A., Grant, P. A., Christie, E. A., Farley, J. R. and Segel, I. H. (1980). Kinetic and chemical properties of ATP sulfurylase from *Penicillium chrysogenum.* In *Sulphur in Biology*, Ciba Foundation Symposium 72, pp. 19–47.

Seubert, P. A., Renosto, F., Knudson, P. and Segel, I. H. (1985). Adenosinetriphosphate sulfurylase from *Penicillium chrysogenum. Arch. Biochem. Biophys.*, **240**, 509–523.

Shaw, W. H. and Anderson, J. W. (1972). Purification, properties and substrate specificity of adenosine triphosphate sulfurylase from spinach leaf tissue. *Biochem. J.*, **127**, 237–247.

Shih, V. E., Abrams, I. F., Johnson, J. L., Carney, M., Mandell, R., Robb, R. M., Cloherty, J. P. and Rajagopalan, K. V. (1977). Sulfite oxidase deficiency. *N. Engl. J. Med.*, **297**, 1022–1028.

Siegel. L. M. (1975). Biochemistry of the sulfur cycle. In D. M. Greenberg (ed.), *Metabolism of Sulfur Compounds*, *Metabolic Pathways*, Vol. VII, 3rd edn., Academic Press, New York, pp. 217–286.

Singer, T. P. (1975). Oxidative metabolism of cysteine and cystine in animal tissues. In D. M. Greenberg (ed.), *Metabolism of Sulfur Compounds*, *Metabolic Pathways*, Vol. VII, 3rd edn., Academic Press, New York, pp. 535–546.

Skarzynski, B., Szczepkowski, T. W. and Weber, M. (1959). Thiosulfate metabolism in the animal organism. *Nature*, **184**, 994–995.

Sörbo, B. (1955). On the catalytic effect of blood serum on the reaction between colloidal sulfur and cyanide. *Acta Chem. Scand.*, **9**, 1656–1660.

Sörbo, B. (1957a). Sulfite and complex-bound cyanide as sulfur acceptors for rhodanese. *Acta Chem. Scand.*, **11**, 628–633.

Sörbo, B. (1957b). Enzymatic transfer of sulfur from mercaptopyruvate to sulfite or sulfinates. *Biochim. Biophys. Acta*, **24**, 324–329.

Sörbo, B. H. (1964). Mechanism of oxidation of inorganic thiosulfate and thiosulfate esters in mammals. *Acta Chem. Scand.*, **18**, 821–823.

Sörbo, B. (1975). Thiosulfate sulfurtransferase and mercaptopyruvate sulfurtransferase. In D. M. Greenberg (ed.), *Metabolism of Sulfur Compounds*, *Metabolic Pathways*, Vol. VII, 3rd edn., Academic Press, New York, pp. 433–456.

Sörbo, B., Hannestad, U., Lundquist, P., Mårtensson, J. and Öhman, S. (1980). Clinical chemistry of mercaptopyruvate and its metabolites. In D. Cavallini, G. E. Gaull and V. Zappia (eds.), *Natural Sulfur Compounds*, Plenum, New York, pp. 463–470.

Stelmaszyńska, T. and Zgliczynski, J. M. (1980). The role of myeloperoxidase in phagocytosis, with special regard to HCN formation. In B. Vennesland, E. E.

Conn, C. J. Knowles, J. Westley and F. Wissing (eds.), *Cyanide in Biology*, Academic Press, London, pp. 371–383.

Stipani, I., Bonvino, V., Schiavulli, N. and Palmieri, F. (1980). Transporto di H^+ nei mitochondri indotto dall'aggiunta di solfito, solfato e tiosulfato. *Boll. Soc. It. Biol. Sper.*, **56**, 1430–1436.

Szczepkowski, T. W. (1961). Udzial rodanazy w metabolicznym powstawaniu tiosiarczanu. *Acta Biochim. Polon.*, **8**, 251–264.

Szczepkowski, T. W. and Wood, J. L. (1967). The cystathionase–rhodanese system. *Biochim. Biophys. Acta*, **139**, 469–478.

Takano, B., McKibben, M. A. and Barnes, H. L. (1984). Liquid chromatographic separation and polarographic determination of aqueous polythionates and thiosulfate. *Anal. Chem.*, **56**, 1594–1600.

Tonzetich, J. and Catherall, D. M. (1976). Metabolism of [^{35}S]-thiosulfate and [^{35}S]-thiocyanate by human saliva and dental plaque. *Arch. Oral Biol.*, **21**, 451–458.

Trudinger, P. A. and Loughlin, R. E. (1981). Metabolism of simple sulphur compounds. In A. Neuberger and L. L. M. Van Deenen (eds.), *Comprehensive Biochemistry*, Vol. 19A, Elsevier, Amsterdam, pp. 165–256.

Tuovinen, O. H. and Nickolas, D. J. D. (1977). Bacterial oxidation of polythionates. *Appl. Environ. Microbiol.*, **33**, 477–479.

Tweedie, J. W. and Segel, I. H. (1971). Adenosine triphosphate sulfurylase from *Penicillium chrysogenum*. *J. Biol. Chem.*, **246**, 2438–2446.

Uhteg, L. and Westley, J. (1979). Purification and steady-state kinetic analysis of yeast thiosulfate reductase. *Arch. Biochem. Biophys.*, **195**, 211–222.

Ullrich, K. J., Rumrich, G. and Klöss, S. (1980). Bidirectional active transport of thiosulfate in the proximal convolution of the rat kidney. *Pfluger's Arch.*, **387**, 127–132.

Volini, M. and Alexander, K. (1981). Multiple forms and multiple functions of the rhodaneses. In B. Vennesland, E. E. Conn, C. J. Knowles, J. Westley and F. Wissing (eds.), *Cyanide in Biology*, Academic Press, London, pp. 77–91.

Weisiger, R. A. and Jakoby, W. B. (1979). Thiol S-methyltransferase from rat liver. *Arch. Biochem. Biophys.*, **196**, 631–637.

Weisiger, R. A. and Jakoby, W. B. (1980). S-Methylation: thiol S-methyltransferase. In W. B. Jakoby (ed.), *Enzymatic Basis of Detoxication*, Vol. II, Academic Press, New York, pp. 131–140.

Westley, A., and Westley, J. (1984). Thin-layer chromatography of thiosulfonate anions. *Anal. Biochem.*, **142**, 163–166.

Westley, J. (1973). Rhodanese. *Advan. Enzymol.*, **39**, 327–368.

Westley, J. (1977). Sulfane-transfer catalysis by enzymes. In E. E. van Tamelen (ed.), *Bioorganic Chemistry*, Vol. I, Academic Press, New York, pp. 371–390.

Westley, J. (1980). Rhodanese and the sulfane pool. In W. B. Jakoby (ed.), *Enzymatic Basis of Detoxication*, Vol. II, Academic Press, New York, pp. 245–262.

Westley, J. (1981a). Cyanide and sulfane sulfur. In B. Vennesland, E. E. Conn, C. J. Knowles, J. Westley and F. Wissing (eds.), *Cyanide in Biology*, Academic Press, London, pp. 61–76.

Westley, J. (1981b). Thiosulfate:cyanide sulfurtransferase (rhodanese). *Methods Enzymol.*, **77**, 285–291.

Westley, J., Adler, H., Westley, L. and Nishida, C. (1983). The sulfurtransferases. *Fundam. Appl. Toxicol.*, **3**, 377–382.

Wood, J. L. (1975). Biochemistry. In A. A. Newman (ed.), *The Chemistry and Biochemistry of Thiocyanic Acid and its Derivatives*, Academic Press, New York, pp. 156–221.

Wood, J. L., Williams, E. F. and Kingsland, N. (1947). The conversion of thiocyanate sulfur to sulfate in the white rat. *J. Biol. Chem.*, **170**, 251–259.

Yokoyama, E., Yoder, R. E. and Frank, N. R. (1971). Distribution of ^{35}S in the blood and its excretion in urine of dogs exposed to $^{35}SO_2$. *Arch. Environ. Health*, **22**, 389–395.

6

Thioethers, thiols, dithioic acids and disulphides: Phase I reactions

L. A. Damani
Chelsea Department of Pharmacy, King's College London, Manresa Road, London SW3 6LX, UK

SUMMARY

1. Thioether (or sulphide) groups occur in a large number of xenobiotic molecules. Alkylaryl sulphides may undergo S-dealkylation to thiols. The more common reaction, however, is the S-oxygenation of thioethers to their corresponding sulphoxides and sulphones. There is recent evidence that thioethers may also be converted into soluble methyl sulphonium ions.
2. Thiol groups occur less frequently in xenobiotic molecules. Oxidation of thiols may afford either disulphides (homodisulphides or mixed disulphides with cysteine, glutathione or protein thiol groups), or S-oxygenated metabolites (sulphenic, sulphinic and sulphonic acids).
3. Dithioic acid groups are only rarely encountered in xenobiotic molecules. Oxidative reactions are minor routes of metabolism for dithioic acids, the majority of the compound undergoing extensive phase II metabolism at the available −SH group.
4. Disulphide groups are present in numerous xenobiotics. Such compounds are readily reduced to thiols. Oxygenation at the sulphur atoms is also possible, but the role of this pathway in the *in vivo* metabolism of disulphides is uncertain.

6.1 INTRODUCTION

Thioether functionalities (−S−) occur extensively in xenobiotics. Thioethers may be ingested accidently as environmental contaminants (e.g. sulphur pesticides), or intentionally as components in food (e.g. dimethyl sulphide in crushed garlic) or as medicinal agents. Additionally, thioethers may be generated *in vivo* by the action of thiol S-methyltransferases on thiols, or by reduction of sulphoxide and possibly sulphones. Historically, the first reported metabolic pathway for sulphur-containing molecules was that of thioether sulphoxidation. During studies of the metabolism of the anthelmintic compund phenothiazine (**1**) in calves, pigs and sheep, phenothiazine sulphoxide (**2**) was identified as a metabolite in several tissues (Clare, 1947). Since this report, S-oxygenation to sulphoxides and sulphones has been recognized as a major route of biotransformation for thioethers (Ziegler, 1982; Mitchell and Waring, 1986; Damani, 1987). Other routes of metabolism for thioethers include S-dealkylation (Mazel *et al.*, 1964) and S-methylation (Hoffman *et al.*, 1988).

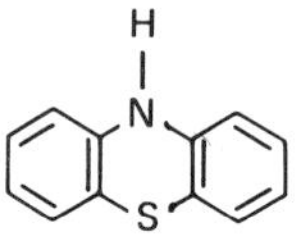

(1) 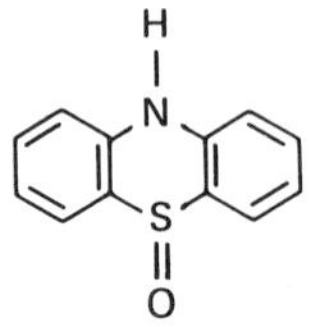   **(2)**

Thiol groups (–SH) occur less frequently in xenobiotic molecules, but they may be formed by the *in vivo* reduction of disulphides, or the S-dealkylation of thioethers. They may also be produced by the action of cysteine conjugate β-lyase enzymes on S-substituted cysteine conjugates (Tateishi *et al.*, 1978). The thiol group is chemically very reactive, and a variety of phase I and II reactions are possible with thiol-containing xenobiotics (Mannervik, 1982). Dithioic acids (see structure below)

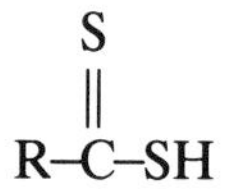

$$R–C–SH$$
$$\overset{\displaystyle S}{\underset{\displaystyle \parallel}{}}$$

are only rarely encountered in xenobiochemistry, and only one such compound has been studied in any detail — diethyldithiocarbamic acid (**3**), the pharmacologically active metabolite of disulfiram or Antabuse® (**4**). The free thiol group is mostly the target of metabolic attack, affording several phase II metabolites, but oxidative pathways may also contribute to the metabolism of diethyldithiocarbamic acid (Ziegler, 1982). Disulphide linkages (–S–S–) are present in numerous constituents in diet (e.g. dimethyl disulphide in onions), and in other foreign sulphur compounds. The main route of metabolism for disulphides probably involves an initial reduction to the corresponding thiol.

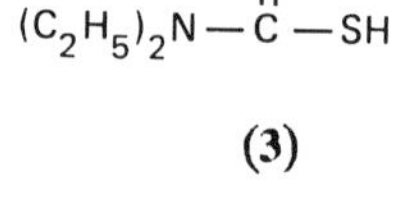

$(C_2H_5)_2N–\overset{S}{\overset{\parallel}{C}}–SH$ $(C_2H_5)_2N–\overset{S}{\overset{\parallel}{C}}–S–S–\overset{S}{\overset{\parallel}{C}}–N(C_2H_5)_2$

(3) 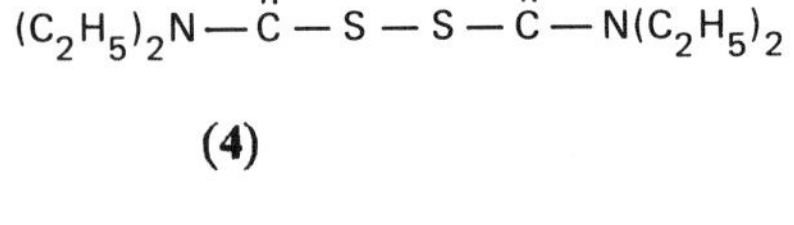   **(4)**

This chapter describes the chemistry of thioethers, thiols, dithioic acids and disulphides which is necessary for an understanding of the fate of these compounds in biological systems. The mechanisms of their phase I metabolic reactions (oxidations, reductions, hydrolyses) are presented, with an outline of the nature of the enzymes catalysing these reactions. Finally, an overview is provided of the pharmacological and toxicological significance of these reactions in mammals. The following chapter in this volume (chapter 7, Volume 1, Part A) deals in detail with phase II (conjugation) reactions of functional groups discussed in this chapter. Detailed discussions of the enzyme systems mediating some of the reactions discussed in this chapter are in Volume 2 (Part A, Chapters 3–5).

6.2   CHEMICAL AND ANALYTICAL ASPECTS

6.2.1   Thioethers and their potential phase I metabolites

Thioethers (or sulphides, R–S–R′) are the sulphur analogues of ethers (R–O–R′). The chemical reactions of sulphides are divided into those taking place directly at the divalent sulphur (e.g. S-oxygenations or S-alkylations) and those occurring at another part of the molecule. The thioether sulphur, because of its unshared pair of electrons, can act as a nucleophile. For example it can carry out a nucleophilic substitution at a saturated carbon, to give sulphonium salts:

$$\begin{array}{l} R \\ \diagdown \\ \ddot{S}: + R''{-}L \rightarrow \\ R' \diagup \end{array} \quad \begin{array}{l} R \\ \diagdown \overset{\oplus}{} \\ S{-}R'' + :\overset{\ominus}{L} \\ R' \diagup \end{array} \qquad (1)$$

(L = leaving group)

The biological formation of methyl sulphonium ion metabolites of thioethers probably utilizes this sulphur nucleophilicity to carry out an S_N2 attack on the methyl group of S-adenosyl-L-methionine (see chapter 6, Volume 1, Part B). The unshared electron pairs of sulphur can also participate in bond formation with oxygen, to give sulphoxides and sulphones.

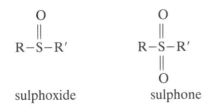

$$\underset{\text{sulphoxide}}{\overset{\displaystyle\overset{O}{\|}}{R{-}S{-}R'}} \qquad\qquad \underset{\text{sulphone}}{\overset{\displaystyle\overset{O}{\|}}{\underset{\displaystyle\underset{O}{\|}}{R{-}S{-}R'}}}$$

These oxidations can be achieved with almost all types of oxidants, e.g. hydrogen peroxide or peracetic acid. Use of one equivalent mole of the oxidant allows the synthesis of sulphoxide, whereas an excess of oxidant is used to carry the reaction through to the sulphone. Even under controlled conditions it is difficult to avoid overoxidation to the sulphone, and some separation technique needs to be utilized to get the sulphoxide in pure form. Sulphoxidation of thioethers results in the decrease of nucleophilicity, but an increase in polar character and hence aqueous solubility. As the oxidation state increases further, the high valency sulphur functional group (sulphone) becomes a good leaving group, and in some cases it can be eliminated by a nucleophilic displacement reaction with glutathione. The sulphoxide function is highly polarized and the oxygen in the S=O group can accept hydrogen bonds; the sulphone function is quite different and is unable to participate in hydrogen bonding.

The geometry of divalent sulphur needs special mention. Sulphur, along with-other second-row atoms, exhibits tetrahedral geometry. When the two substituents on the divalent sulphur of thioethers are dissimilar, the lone pairs on the sulphur atom become prochiral, and the sulphoxides produced exist as enantiomers. Sulphoxides undergo stereomutation only under extreme conditions, and can therefore be resolved into their enantiomers. Enzyme-mediated sulphoxidations are often stereospecific or stereoselective, and analytical methods are now available for the selective estimation of these enantiomers (see chapter 4, Volume 2, Part A). Most thioethers and their oxidation products are readily isolated from biological media by solvent extraction and analysed by simple chromatographic techniques (gas liquid chromatography (GLC); high performance liquid chromatoghraphy (HPLC) (Cates and Meloan, 1963; Cox and Przyjazny, 1977). The methyl sulphonium metabolities are highly polar and non-volatile, and their analysis requires cation exchange HPLC.

6.2.2   Thiols and their potential phase I metabolites

Thiols (R−SH) are sulphur analogues of alcohols (R−OH); they are also known as thioalcohols or mercaptans. Many thiols have very low boiling points and objectionable odours. For example the well-known scent of the skunk is due to n-butyl thiol. The thiol group (−SH) is very reactive, and its chemistry has been discussed in detail elsewhere (Patai, 1974). The acidity of the thiol group is important in its chemistry, since the ionized form, i.e. the thiolate ion −S$^-$, is the reactive species in the reaction of thiols. Most aliphatic thiols and aliphatic aminothiols have a pK_a of about 8 and above, whereas aromatic thiols have somewhat lower pK_a values ($\sim$6–7). Despite the apparent similarity between alcohols and thiols, the latter are much stronger acids (the S−H bond is much weaker than the O−H bond). As in the case of the thioether sulphur, the thiol sulphur is also a powerful nucleophile and can participate in a nucleophilic substitution reaction at a saturated carbon to give a disulphide:

$$R'-\overset{..}{S}H + R-L \rightarrow R'-S-R + H-L \tag{2}$$

(L = leaving group)

This high reactivity of thiols means that they are very readily oxidized to disulphides in the environment, thus limiting the amount of free thiols entering organisms. The disulphides may be reduced within cells to regenerate the thiols.

Most moderately strong oxidizing agents, e.g. KMnO$_4$ or H$_2$O$_2$, will readily oxidize thiols to their sulphonic acid (R−SO$_3$H) derivatives. This involves three two-electron oxidation steps, but the intermediate sulphenic (R−SOH) and sulphinic (R−SO$_2$H) acids are difficult to obtain even under carefully controlled conditions. Sulphenic acids are unstable with respect to disproportionation (self-oxidation/reduction) and cannot usually be isolated (Hendrickson *et al.*, 1970). The sulphinic acids may be prepared by zinc reduction of sulphonyl chlorides. The disulphide derivatives of thiols are usually prepared by treatment with mild oxidizing agents (e.g. iodine). The analytical aspects of thiol metabolism present special problems, in view of the high reactivity of the substrate and most of the potential metabolites. These aspects are discussed in more detail in Volume 2 (chapter 2, Part B) of this series.

6.2.3 Dithioic acids and their potential phase I metabolites

Dithioic acids are the sulphur analogues of carboxylic acids (see below).

$$
\begin{array}{cc}
S & O \\
\parallel & \parallel \\
R-C-SH & R-C-OH \\
\text{dithioic acid} & \text{carboxylic acid}
\end{array}
$$

This class of chemicals was amongst the earliest discovered in organosulphur chemistry. For example, Debus wrote in *Liebig's Annalen der Chemie* on the synthesis of dithiocarbamic acid in 1850; this compound is the sulpur analogue of carbamic acid (see below).

$$
\begin{array}{cc}
S & O \\
\parallel & \parallel \\
H_2N-C-SH & H_2N-C-OH \\
\text{dithiocarbamic acid} & \text{carbamic acid}
\end{array}
$$

A large number of dithiocarbamates and related compounds have since been prepared and many have found applications in agriculture and medicine (Thorn and Ludwig, 1962). Since very few free dithioic acids have been studied in as much detail, the rest of the discussion will be focused on dithiocarbamic acid or dithiocarbamates (see below).

$$
\begin{array}{cc}
S & S \\
\parallel & \parallel \\
R^1R^2N-C-SH & R^1R^2N-C-S^-Na^+ \\
\text{dithiocarbamic acids} & \text{dithocarbamates}
\end{array}
$$

Dithiocarbamic acids are readily oxidized by a variety of mild oxidising agents (e.g. iodine, hydrogen peroxide) to *thiuram disulphides* (Braun, 1902):

$$
2(R^1R^2N-\overset{\displaystyle S}{\overset{\displaystyle \parallel}{C}}-S^-Na^+) + I_2 \rightarrow R^1R^2N-\overset{\displaystyle S}{\overset{\displaystyle \parallel}{C}}-S-S-\overset{\displaystyle S}{\overset{\displaystyle \parallel}{C}}-NR^1R^2 + 2NaI \quad (3)
$$

This reaction is important because the commercially used compounds (e.g. disulfiram or Antabuse®) are mostly thiuram disulphides. These disulphides can be reduced with zinc dust to the dithiocarbamates. The dithiocarbamic acids can conveniently be converted into their alkyl esters by reaction with alkyl halides. These dithiocarbamate esters are considerably more stable towards acids and alkaline reagents than are the carboxylate esters. It is noteworthy here that such methyl esters of dithiocarbamic acids have been reported as stable urinary metabolites after the administration of the thiuram drug disulfiram.

6.2.4   Disulphides and their potential phase I metabolites

Disulphides $(R-S-S-R')$ are sulphur analogues of peroxides $(R-O-O-H)$. Disulphides can be reduced to thiols by a variety of reactions, e.g. reaction with a thiol:

$$R-S-S-R' + XS^- \rightarrow R-S-S-X + R'S^- \tag{4}$$

$$R-S-S-X + XS^- \rightarrow X-S-S-X + RS^- \tag{5}$$

Disulphides may also be reduced by direct chemical reduction with sodium borohydride, which is probably more suitable for small molecular weight disulphides. Electrochemical reduction of disulphides has also been described (Kadin, 1987). Disulphides may be oxidized to thiosulphenates (thiosulphoxides) which hydrolyse to the sulphinic acid and thiol. Thiosulphinates (thiosulphones) may also be produced by further oxidation of thiosulphenates; the thiosulphinates hydrolyse to sulphenic and sulphinic acids (see Ziegler, 1980). As in the case of thiols (see section 6.2.2), analysis of trace amounts of xenobiotic disulphides in biological fluids presents numerous problems. Some of these aspects are covered in detail in chapter 2, Volume 2, Part A, of this series.

6.3   BIOCHEMICAL ASPECTS

6.3.1   Phase I metabolism of thioethers

The metabolism of a variety of thioethers (sulphides) has been investigated since the early report of phenothiazine sulphoxidation (see section 6.1). The types of metabolic reactions possible are summarized below:

$$R-S-R' \rightarrow R-\overset{\overset{\displaystyle O}{\|}}{S}-R' \rightarrow R-\overset{\overset{\displaystyle O}{\|}}{\underset{\underset{\displaystyle O}{\|}}{S}}-R' \tag{6}$$

sulphide   sulphoxide   sulphone

$$R-S-R' \rightarrow R-\overset{+}{\underset{\underset{\displaystyle CH_3}{|}}{S}}-R' \tag{7}$$

sulphide   sulphonium ion

$$R-S-R' \rightarrow R-SH + R'CHO \tag{8}$$

sulphide   thiol   aldehyde

$(R' = $ alkyl group$)$

Most alkyl, alkylaryl and aryl sulphides are nucleophilic, and almost without exception they are metabolized to sulphoxide and sulphone metabolites (eq. (6)), both *in vitro* and *in vivo* (Ziegler, 1980, 1982; Mitchell and Waring, 1986; Damani, 1987).

6.3.1.1 Dialkyl sulphides

An example of a simple dialkyl sulphide affording *S*-oxygenated metabolites is dimethyl sulphide. Williams *et al.* (1966) administered about 3.4 g (1.4 g/kg) subcutaneously in four divided doses to rabbits and collected urine for six days. About 20% of the dose was excreted as the sulphoxide and a further 10% as the sulphone. It is interesting that dimethyl sulphide had been studied previously (Maw, 1953) in the rat; administration orally or by injection did not lead to an increase in total 'ethereal sulphate'. The sulphur in dimethyl sulphide is clearly not completely oxidizied *in vivo* to sulphate, but its intermediate oxidation products are the elimination products in urine. Dimethyl sulphide is not a substrate *in vitro* for microsomal *S*-demethylating enzymes (Mazel *et al.*, 1964), and as a general rule dialkyl sulphides do not undergo *S*-dealkylation to any significant extent. Administration of dimethyl sulphoxide to man and animals (Hucker *et al.*, 1966, 1967; Layman and Jacob, 1985) resulted in the almost total recovery of the material, either as unchanged drug (60–68%) or as the sulphone metabolite (16–21%). Administration of the sulphone did not result in the appearance of either dimethyl sulphoxide or dimethyl sulphide. A number of other dialkyl sulphides have been investigated, and the following simplified chemical expression is the summary of these cumulative observations:

$$\underset{}{R-S-R'} \rightleftharpoons \overset{\overset{\textstyle O}{\|}}{R-S-R'} \rightarrow \underset{\underset{\textstyle O}{\|}}{\overset{\overset{\textstyle O}{\|}}{R-S-R'}} \tag{9}$$

This implies that sulphides and sulphoxides can undergo redox interconversions *in vivo*, but the sulphone, once formed, is not reduced back to the sulphoxide or sulphide (see chapter 5, Volume 1, Part B, for a further discussion of sulphoxide–sulphone reduction).

S-Dealkylation is usually a very minor route of metabolism for dialkyl sulphides, if it occurs at all. Hanzlik (1984) has offered an explanation as to why *S*-dealkylations only occur infrequently, compared with *O*- and *N*-dealkylations. However, more detailed studies are required to elucidate mechanisms, since the dealkylations are mediated by cytochrome P-450, whereas most sulphoxidations at nucleophilic sulphur are mediated by the flavin-containing monooxygenase.

Thioethers, e.g. dimethyl sulphide, diethyl sulphide etc., have also recently been demonstrated to afford methyl sulphonium ions *in vitro* and *in vivo* (Hoffman *et al.*, 1988; Mozier *et al.*, 1988). This reaction can be viewed as an 'oxidation' since there is a net increase in charge at the heteroatom. The *in vitro* data clearly show that the methyl group is derived from *S*-adenosyl-L-methionine, and this transfer is effected by cytosolic thioether *S*-methyltransferases (TEMTase) present in almost all mammals tested to date.

6.3.1.2 Alkylaryl sulphides

S-Dealkylation is a known route of metabolism for alkylaryl sulphides. This reaction proceeds in a similar fashion to *O*- and *N*-dealkylations, i.e. via an initial α-carbon hydroxylation of the alkyl group to form an unstable α-hydroxy compound, which rearranges by carbon–sulphur bond fission to the thiol and the corresponding aldehyde:

$$Ar-S-CH_2R' \rightarrow \left[\begin{array}{c} Ar-S-CHR' \\ | \\ OH \end{array} \right] \rightarrow Ar-SH + R'-CHO \tag{10}$$

6-Methylthiopurine (**5**) is an example of a compound that undergoes *S*-demethylation *in vitro* (Mazel *et al.*, 1964). However, 6-mercaptopurine (**6**) is a substrate for *S*-methyltransferases (see chapter 1, Volume 3, Part B), and it is likely that *in vivo* *S*-dealkylation plays an insignificant role in the elimination of alkylaryl sulphides. Whether alkylaryl sulphides are substrates for the TEMTase (see section 6.3.1.1 above) is as yet unclear; Hoffman *et al.* (1988) and Mozier *et al.* (1988) have only tested simple dialkyl sulphides in their preliminary studies.

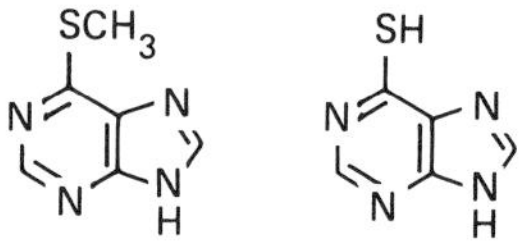

(5)   (6)

By far the most important reactions with alkylaryl sulphides are those that lead to the formation of sulphoxides and sulphones. Since the early report of sulphoxidation of *p*-methylthiolaniline (**7**), several others (e.g. 4-chlorophenylmethyl sulphide, **8**) have been reported to afford sulphoxides and sulphones as major metabolic products (Rose and Spinks, 1948; Oehler and Ivie, 1983). The sulphur atom in alkylaryl sulphides is prochiral (or enantiotopic) and the sulphoxide metabolites exist as enantiomers. Studies to date on mammalian chiral sulphoxidations have almost invariably utilized alkylaryl sulphides (e.g. *p*-tolylethylsulphide, **9**). When this compound was incubated with pig liver flavin-containing monooxygenase, the sulphoxide formed had the *R*-configuration to the extent of 95%. On the other hand, two purified P-450 isozymes from rat liver selectively gave rise to the *S*-sulphoxide (~80% enantiomeric excess) (Waxman *et al.*, 1982).

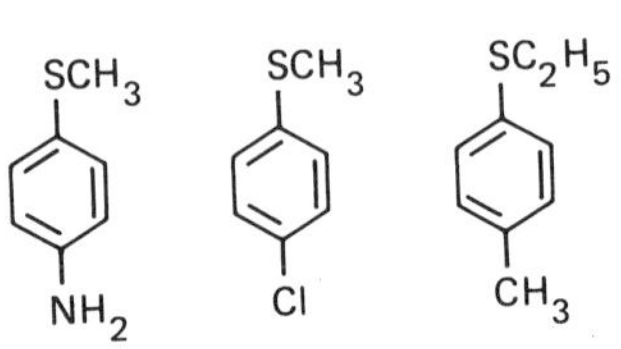

(7)   (8)   (9)

The data from Waxman *et al.* (1982) put into question much of the early work on chiral sulphoxidation which utilized alkylaryl sulphides and crude enzyme preparations. Data interpretation and formulation of stereochemical rules from such studies are difficult on two counts. Firstly it now appears that whereas the more nucleophilic dialkyl sulphides are exclusively *S*-oxygenated by the flavin-containing monooxygenase (Ziegler, 1980, 1982; Houdi and Damani, 1984), the alkylaryl sulphides are substrates for both the flavin and cytochrome P-450 systems (Waxman *et al.*, 1982); the less nucleophilic diaryl sulphides (e.g. dibenzthiophen) on the other hand are only *S*-oxygenated by P-450 (Houdi and Damani, 1984). Secondly the ready interconversion of the redox states (eq. (9)) in microsomal systems has not always been appreciated. The net stereochemical outcome therefore may be due not only to the contribution of at least two different monooxygenases but possibly also to selective chiral processing of the sulphoxide metabolite in crude systems. A judicious choice of substrates and test systems (purified enzymes) is required in future studies, if the data are to be used for the formulation of a set of stereoselectivity rules for enzymic chiral sulphoxidations and for chiral sulphoxide oxidations–reductions. A full account of chiral sulphoxidations is in Volume 2, Part A, chapter 4, of this series.

6.3.1.3 *Diaryl sulphides*
Apart from the early report on the *in vitro* sulphoxidation of 4,4′-diaminodiphenyl sulphide (**10**) by Gillette and Kamin (1960), most of the other examples are where the sulphur is in an aromatic ring, e.g. in phenothiazine (**1**) or in its derivative, thioridazine (**11**). Dibenzthiophen (**12**) and the benzthiophen derivative, Mobam (**13**) are examples of simple aromatic sulphur heterocycles that afford sulphoxides as metabolites (Houdi, 1986; Robin *et al.*, 1970), albeit as minor components. These benzthiophens appear to be extensively metabolized by ring carbon hydroxylation and glutathione conjugation (Bray *et al.*, 1971).

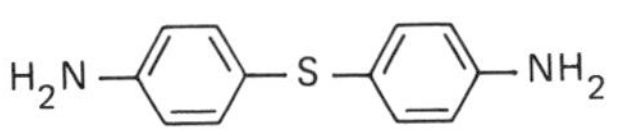

(10) 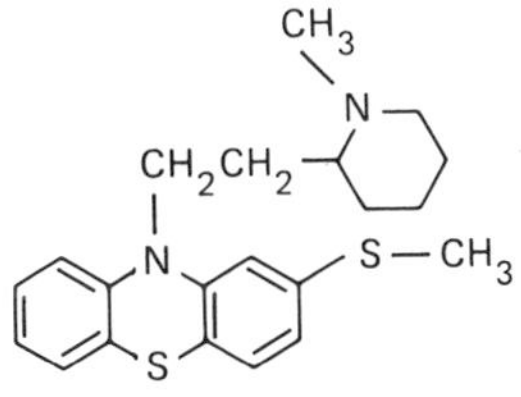   (11)

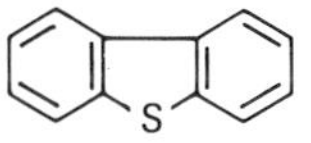

(12) 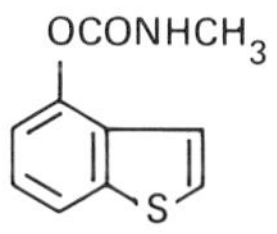   (13)

The phenothiazine derivative, thioridazine (**11**), is an interesting molecule, since it contains two sulphur centres, an aromatic heterocyclic sulphur and an alkylaryl side-chain sulphur, both of which can potentially be oxidized to sulphoxides and sulphones. Ring *S*-oxidation, which is quantitatively an important route of metabolism, leads to pharmacological inactivation, in common with most other phenothiazines (Papadopoulos and Crammer, 1986). Side-chain *S*-oxidation is also important in man, and leads to products that are pharmacologically active; the side-chain sulphoxide (mesoridazine) and sulphone (sulphoridazine) have been demonstrated to possess significant antipyschotic effects in man (Axelsson, 1977). Whereas the antipsychotic activity of the sulphoxide might have been explained by a possible *in vivo* reduction to the sulphide (as in the case of sulindac and sulphinpyrazone), the activity of the sulphone (sulphoridazine) is unlikely to be due to this mechanism, sulphones being metabolically resistant to reduction. The side-chain sulphoxidation products may therefore possess intrinsic pharmacological activity, the initial *S*-oxidation therefore being a 'bioactivation' pathway.

6.3.1.4 *Enzymology of sulphoxidation*
Sulphoxide and sulphone metabolies are hydrophilic and usually chemically stable, and they are therefore readily detected in the urine of animals treated with sulphide drugs. The amount of sulphone formed *in vivo* or *in vitro* is usually less than that of the sulphoxide. This is probably due to the water solubility of the sulphoxides, which presumably limits their partitioning into the catalytic sites on the microsomal monooxygenases. The first oxidation at sulphur in thioethers, i.e. sulphoxide formation, is reversible, but the subsequent reaction to sulphones is irreversible. At least three different enzyme systems have been reported to catalyse sulphoxide formation, cytochrome P-450, the flavin-containing monooxygenase and the microsomal prostaglandin synthetase. Any particular compound may be a substrate for more than one enzyme, a major determinant being the electromolecular environment in which the sulphur occurs (Hunt *et al.*, 1982; Damani and Houdi, 1988). For example, the more nucleophilic divalent sulphur atoms in aliphatic and alicyclic sulphides (e.g. diethylsulphide and tetrahydrothiophen respectively) are oxidized exclusively by the flavin monooxygenase. The sulphur in aromatic heterocyclic rings (e.g dibenzothiophen), where there is partial delocalization due to the aromatic rings, is oxidized exclusively by cytochrome P-450 (Houdi and Damani, 1984; Houdi, 1986). It is of interest that *p*-tolylethylsulphide, an alkylaryl sulphide, is oxidized to a chiral sulphoxide by both the monooxygenases (Waxman *et al.*, 1982). The more nucleophilic sulphides can also function as cosubstrate reductants for the endoperoxide precursors of prostaglandins and thromboxanes, during the prostaglandin synthetase mediated reaction (Egan *et al.*, 1979). However, the *in vivo* significance of this type of *S*-oxygenation for the thioether drugs is unclear. Indeed, most of the mechanistic–enzymology studies have been carried out *in vitro* with model thioethers. The contribution of the various *S*-oxygenases to sulphoxidation of drugs and foreign sulphides *in vivo* is unknown.

6.3.2   Phase I metabolism of thiols and disulphides
The types of phase I metabolic reactions possible at thiol and disulphide functionalities are summarized below:

$$R-SH \rightarrow R-SOH \rightarrow R-SO_2H \rightarrow R-SO_3H \tag{11}$$

$$R-SH \rightarrow R-SOH \xrightarrow{RSH} R-S-S-R + H_2O \tag{12}$$

$$R-S-S-R \rightarrow R-S-\overset{\overset{\displaystyle O}{\|}}{S}-R \xrightarrow{H_2O} RSO_2H + RSH \tag{13}$$

$$R-S-S-R \rightarrow R-S-\overset{\overset{\displaystyle O}{\|}}{S}-R \rightarrow R-\overset{\overset{\displaystyle O}{\|}}{S}-\overset{\overset{\displaystyle O}{\|}}{S}-R \xrightarrow{H_2O} RSO_2H + RSOH \tag{14}$$

Reactivity of the thiols is due to the fact that most thiols are readily ionized at physiological pH to the nucleophilic thiolate anion (see section 6.2.2). Enzymic oxygenation of thiols affords reactive sulphenic acids ($R-SOH$) and then sulphinic acids ($R-SO_2H$) (eq. (11)). The antithyroid drug methimazole (**14**) is metabolized by these pathways *in vitro*, although the intermediate sulphenic acids are not directly demonstrable (Poulsen *et al.*, 1979). The sulphenic acids can react with excess thiols to afford disulphides (eq. (12)). Exogenous disulphides, or those generated *in vivo*, can either be reduced back to thiols or converted into sulphenic and sulphinic acids via the intermediate thiosulphenic and thiosulphinic acids (eqns (13) and (14)). It is not clear how many thiol drugs and other foreign thiols actually undergo the elegant sequences of reactions proposed by Ziegler (1984). Oxidation to the highest oxidation state ($+4$) is not often seen with thiols and pyridine-2-thiol-N-oxide (**15**) is an interesting example. This aryl mercaptan, used as a scalp antiseptic in man, gave rise to the corresponding sulphonic acid as the major metabolite in rats after dermal adminstration (Min *et al.*, 1970); the disulphide was also present in smaller amounts.

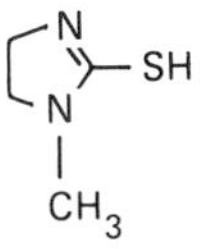

(14) 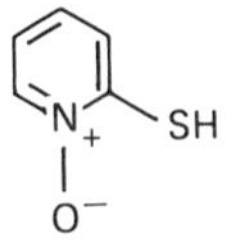   (15)

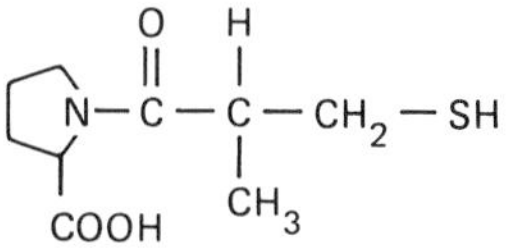

(16) 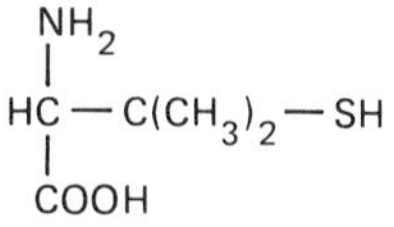   (17)

The thiol drugs captopril (**16**) and penicillamine (**17**) afford disulphides as the main metabolites. In both these drugs, the free thiol group is the only functional group that undergoes biotransformation, to disulphides, mixed disulphides with glutathione or other endogenous mercaptans, and drug-protein disulphides. The thiol drugs and their disulphides probably undergo ready interconversions, but even the formation of these relatively short-lived drug–plasma protein conjugates may be of relevance to the toxicity of these thiols (Crawhall *et al.*, 1979; Park and Yeung, 1983) (also see chapters 2 and 3, Volume 3, Part B).

Conjugation reactions of thiol drugs are probably more relevant to their detoxication and removal from the body, and these are discussed fully in chapter 7 of this volume.

6.3.3   Phase I metabolism of dithioic acids

There have been no systematic investigations on dithioic acid metabolism, and the only data available are for diethyldithiocarbamic acid (**3**). Disulfiram (**4**), or tetraethylthiuram disulphide, renders humans sensitive to ethyl alcohol (Hald and Jacobsen, 1948) and has therefore been used in the treatment of chronic alcoholism under the trade name Antabuse®, Abstinyl® or Disulphiram®. This thiuram disulphide is reduced *in vivo* to diethyldithiocarbamic acid (**3**), the pharmacologically active metabolite. The metabolism of **3** occurs predominately at the dithioic acid functionality, i.e. at $-C(S)SH$. The major routes of metabolism are conjugation with glucuronic acid and methyl group (see chapter 7, Volume 1, Part A). *S*-Methylation of **3** to the *S*-methylthiocarbamate (**18**) is of interest, since the molecule can now undergo phase I oxidative desulphuration to the thiocarbamate (**19**), the released sulphur presumably being oxidized to sulphate and excreted in urine. Administration of **18** to rats resulted in extensive desulphuration ($>80\%$ of dose) (Gessner and Jakabowski, 1972). The thiocarbamide (**19**) can itself undergo oxidative reactions which ultimately result in the loss of the remaining sulphur and its appearance as sulphate in urine. Other oxidative pathways undoubtedly exist, since carbon disulphide and diethylamine have also been found as *in vitro* and *in vivo* metabolites.

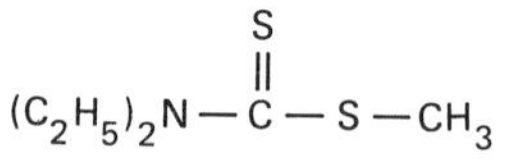

$$(C_2H_5)_2N-\overset{\overset{\displaystyle S}{\|}}{C}-S-CH_3 \qquad (C_2H_5)_2N-\overset{\overset{\displaystyle O}{\|}}{C}-S-CH_3$$

$$\textbf{(18)} \qquad\qquad\qquad \textbf{(19)}$$

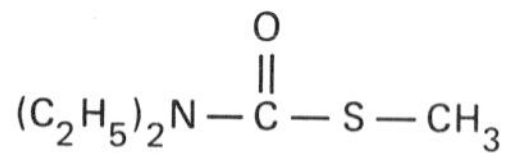

Taylor and Ziegler (1987) have recently reported an interesting structure-activity study on a series of dithioic acids. The flavin-containing monooxygenase can apparently carry out the S-oxygenation of certain dithioic acids to mono- and dioxygenated products (equivalent to sulphenic and sulphinic acids) (see Eq. 15)

$$R-\overset{\overset{\displaystyle S}{\|}}{C}-SH \rightarrow R-\overset{\overset{\displaystyle S}{\|}}{C}-SOH \rightarrow R-\overset{\overset{\displaystyle S}{\|}}{C}-SO_2H \qquad (15)$$

The proposed sequence of oxidation of dithioic acids has, however, not been unequivocally established, but has been inferred from reaction stoichiometry studies (Taylor and Ziegler, 1987; Ziegler, 1988). Amongst the dithioic acids that are substrates for the flavin enzyme include N,N-diethyldithiocarbamic acid (**3**), 4-dimethylaminodithiobenzoic acid dithiobenzoic acid and dithiosalicyclic acid. It is not known yet whether such S-oxygenations occur in the *in vivo* metabolism of dithioic acids.

6.4 PHARMACOLOGICAL AND TOXICOLOGICAL ASPECTS

Sulphoxides and sulphones are not usually chemically reactive, and therefore do not pose any toxicological problems. The sulphoxides and sulphones of some types of thioethers may be good leaving groups and participate in reaction with glutathione. The thiocarbamate herbicide EPTC (**20**) and alkylthiotriazine herbicide cyanatryn (**21**) are good examples of this sulphoxide reactivity (Hutson, 1981). EPTC is metabolically converted to the thiocarbamate sulphoxide, an active carbamoylating agent. Cyanatryn is also metabolized to a sulphoxide, a powerful triazinylating agent which shows a particular reactivity towards thiol groups. Therefore one needs to keep this aspect in mind in designing drugs for use in man.

(**20**)   (**21**)

In many cases sulphoxidation (e.g. chloropromazine, cimetidine) results in the decrease of pharmacological activity. In other cases, sulphoxidation may actually impart activity (e.g. with thioridazine; see section 6.3.1.3). Where the intrinsic activity resides with a chiral sulphoxide molecule, any alterations in the enantiomeric composition of the administered drug may alter pharmacological activity. It is likely that when synthetic racemic sulphoxides are used as drugs, or when sulphoxides are produced as metabolites of prochiral thioethers, the two enantiomers may possess different biological properties. Futile sulphide–sulphoxide cycling *in vivo* has been demonstrated for sulindac (a sulphoxide) and for metiamide (a thioether). These enzymatic redox interconversions would be expected to alter the enantiomeric composition of the sulphoxides, with possible concomitant changes in their pharmacological or toxicological properties. This aspect of sulphur xenobiochemistry has not yet been studied, but should now be looked at in view of the availability of chiral HPLC columns for the separation of enantiomeric sulphoxides.

Finally, it will have become apparent from the preceding sections that thioethers, thiols, dithioic acids and disulphides undergo numerous phase I metabolic reactions. In the case of thioethers, sulphoxidation is the route which leads to products that are eliminated in urine, i.e. phase II metabolism is not required. In the other compounds, i.e. thiols, dithioic acids and disulphides, conjugation reactions are the main detoxication pathways that lead ultimately to the formation of hydrophilic metabolites that can be eliminated. These phase II reacitons are discussed in the following chapter (chapter 7, Volume 1, Part A).

REFERENCES

Axelsson, R. (1977). On the serum concentrations and antipsychotic effects of thioridazine, thioridazine side-chain sulfoxide and thioridazine side-chain sulphone, in chronic psychotic patients. *Current Therap. Res.*, **21**, 587–605.

Braun, V. (1902). Thiuram disulphides and isothiuram disulphides. I. *Ber.*, **35**, 817.

Bray, H. G., Carpanini, F. M. B., and Waters, B. D. (1971). The metabolism of thiophen in the rabbit and rat. *Xenobiotica*, **1**, 157–168.

Cates, V. E. and Meloan, C. E. (1963). Separation of sulphones by gas chromatography. *J. Chromatogr.*, **11**, 472–478.

Clare, N. T. (1947). A photoosensitized keratitis in young cattle following the use of phenothiazine as an anthelmintic: the metabolism of phenothiazine in ruminants. *Aust. Vet. J.*, **23**, 340–344.

Cox, J. A. and Przyjazny, A. (1977). High pressure liquid chromatography of selected sulphur compounds. *Anal. Lett.*, **10**, 869–885.

Crawhall, J. C., Lecavalier, D. and Ryan, P. (1979). Penicillamine, its metabolism and therapeutic applications. a review. *Biopharmaceutics Drug Disposit.*, **1**, 73–95.

Damani, L. A. (1987). Metabolism of sulphur-containing drugs, in D. J. Benford, J. W. Bridges and G. G. Gibson (eds.), *Drug Metabolism — From Microbes to Man*, Taylor & Francis, London, pp. 581–603.

Damani, L. A. and Houdi, A. A. (1988). Cytochrome P-450 and FAD-monooxygenase mediated S- and N-oxygenations. *Drug Metab. Drug Interact.*, **6**, 350–363.

Debus, H. (1850). Über die Verbindungen der sulfocarbaminsäure. *Ann. Chem. [Liebigs)*, **73**, 26–30.

Egan, R. W., Gale, P. H. and Kuehl, F. A. Jr. (1979). Reduction of hydroperoxides in prostaglandin biosynthetic pathway by microsomal peroxidase. *J. Biol. Chem.*, **254**, 3295–3302.

Gessner, T. and Jakubowski, M. (1972). Diethyldithiocarbamic acid methyl ester: a metabolite of disulfiram. *Biochem. Pharmacol.*, **21**, 219–230.

Gillette, J. R. and Kamin, J. J. (1960). The enzymatic formation of sulfoxides: the oxidation of chlorpromazine and 4,4'-diaminodiphenyl sulfide by guinea pig liver microsomes. *J. Pharmacol. Exp. Ther.*, **130**, 262–267.

Hald, J. and Jacobsen, E. (1948). A drug sensitizing the organism to ethyl alcohol. *Lancet*, **255**, 1001–1004.

Hanzlik, R. P. (1984). Prediction of metabolic pathways — sulphur functional groups. In J. Caldwell and G. D. Paulson (eds.), *Foreign Compound Metabolism*, Taylor & Francis, London, pp. 65–78.

Hendrickson, J. B., Cram, D. J. and Hammond, G. S. (1970). *Organic Chemistry*, 3rd edn., McGraw-Hill, New York, pp. 789–814.

Hoffman, J. L., Mozier, N. M. and Warner, D. R. (1988). S-Adenosylmethionine: Thioether S-Methyltransferase (TEMTase) in the metabolism of sulphur xenobiotics. *First Int. Symp. Sulp. Xenobiochemistry*, September 1988, London, Abstract No. 23.

Houdi, A. A. (1986). Studies on metabolic sulphoxidation of alkyl and aryl thioethers- role of cytochrome P-450 and FAD-containing monooxygenases. *PhD Thesis*, University of Manchester.

Houdi, A. A. and Damani, L. A. (1984). Cytochrome P-450 and non-cytochrome P-450 sulphoxidations. *J. Pharm. Pharmacol.*, **36**, (suppl.), 62P.

Hucker, H. B., Ahmad, P. M. and Miller, J. K. (1966). Absorption, distribution and metabolism of dimethylsulphoxide int he rat, rabbit and guinea pig. *J. Pharmacol. Exp. Ther.*, **154**, 176–184.

Hucker, H. B., Miller, J. K., Hochberg, A., Brobyn, R. D., Riordan, F. H. and Calesnick, B. (1967). Studies on the absorption, excretion and metabolism of dimethylsulphoxide (DMSO) in man. *J. Pharmacol. Exp. Ther.*, **155**, 309–317.

Hunt, P. A., Mitchell, S. C. and Waring, R. H. (1982). Some properties of sulphoxidising enzymes. In R. Snyder, D. V. Parke, J. J. Kocsis, D. J. Jollow, G. G. Gibson and C. M. Whitmer (eds.), *Biological Reactive Intermediates II — Chemical Mechansims and Biological Effects*, Plenum, New York, pp. 1255–1262.

Hutson, D. H. (1981) S-Oxygenation in herbicide metabolism in mammals. In J. D. Rosen, P. S. Magee and J. E. Casida (eds.), *Sulphur in Pesticide Action and Metabolism*, American Chemical Society, Washington, pp. 53–64.

Kadin, H. (1987). Electrochemical reduction of disulphides. In W. B. Jakoby and O. W. Griffith (eds.), *Methods in Enzymology*, Vol. 143, *Sulphur and Sulphur Amino Acids*, Academic Press. London, pp. 257–264.

Layman, D. L. and Jacob, S. W. (1985). Absorption, metabolism and excretion of diemthyl sulphoxide by Rhesus monkeys. *Life Sci.*, **37**, 2431–2437.

Mannervik, B. (1982). Mercaptans. In W. B. Jakoby, J. R. Bend and J. Caldwell (eds.), *Metabolic Basis of Detoxication*, Academic Press, New York, pp. 185–206.

Maw, G. A. (1953). Observations on the fate of some aliphatic sulphonic acids in the rat. *Biochem. J.*, **55**, 37–41.

Mazel, P., Henderson, J. F. and Axelrod, J. (1964). S-Demethylation by microsomal enzymes. *J. Pharmacol. Exp. Ther.*, **143**, 1–6.

Min, B. H., Parekh, C., Goldberg, L. and McChesney, E. W. (1970). Experimental studies of sodiumpyridinethione: II. Urinary excretion following topical application to rats and monkeys. *Food Cosmetics Toxicol.*, **8**, 161–166.

Mitchell, S. C. and Waring, R. H. (1986). The early history of xenobiotic sulfoxidation. *Drug Metabs. Rev.*, **16**, 255–284.

Mozier, N. M., McConnell, K. P. and Hoffman, J. L. (1988). S-Adenoxyl-L-methionine: Thioether S-Methyltransferase, a new enzyme in sulphur and selenium metabolism. *J. Biol. Chem.*, **263**, 4527–4531.

Oehler, D. D. and Ivie, G. W. (1983). Metabolism of 4-chlorophenyl methyl sulphide and its sulphone analogue in cattle and sheep. *Arch. Environ. Contam. Toxicol.*, **12**, 227–233.

Papadopoulos, A. S. and Crammer, J. L. (1986). Sulphoxide metabolites of thioridazine in man. *Xenobiotica*, **16**, 1097–1107.

Park, B. K. and Yeung, J. H. K. (1983). Disposition of captopril in the rat. In S. C. Mitchell and R. H. Waring (eds.), *Sulphur in Xenobiotics*, Birmingham University Press, pp. 97–100.

Patai, S. (ed.) (1974). *The Chemistry of the Thiol Group*, Wiley, New York.

Poulsen, L. L., Hyslop, R. M. and Ziegler, D. M. (1979). S-Oxygenation of N-substituted thioureas catalysed by the pig liver microsomal FAD-containing monooxygenase. *Arch. Biochem. Biophys.*, **198,** 78–88.

Robbin, J. D., Bakke, J. V. and Feil, V. J. (1970). Metabolism of benzo(b)thien-4-yl methyl-carbamate (Mobam) in diary goats and a lactating cow. *J. Agric. Food Chem.*, **18,** 130–134.

Rose, F. L. and Spinks, A. (1948). Metabolism of aryl sulphides: Part I. Conversion of *p*-methylthioaniline to *p*-methylsulphonylaniline in the mouse. *Biochem. J.*, **43,** vii.

Tateishi, M., Suzuki, S. and Shimizu, H. (1978). Cysteine conjugate β-lyase in rat liver: a novel enzyme catalyzing formation of thiol-containing metabolites of drugs. *J. Biol. Chem.*, **253,** 8854–8859.

Taylor, K. L. and Ziegler, D. M. (1987). Studies on substrate specificity of the hog liver flavin-containing monooxygenase. Anionic organic sulfur compounds. *Biochem. Pharmacol.*, **36,** 141–146.

Thorn, G. D. and Ludwig, R. A. (1962). *The Dithiocarbamates and Related Compounds*, Elsevier, Amsterdam.

Waxman, D. J., Light, D. R. and Walsh, C. (1982). Chiral sulphoxidations catalysed by rat liver cytochromes P-450. *Biochemistry*, **21,** 2499–2507.

Williams, K. I. H., Burnstein, S. H. and Layne, D. S. (1966). Metabolism of dimethyl sulphide, dimethyl sulphoxide and dimethyl sulphone. *Arch. Biochem. Biophys.*, **117,** 84–87.

Ziegler, D. M. (1980). Microsomal flavin-containing monooxygenase: Oxygenation of nucleophilic nitrogen and sulphur compounds. In W. B. Jakoby (ed.), *Enzymatic Basis of Detoxication*, Academic Press, New york, pp. 201–227.

Ziegler, D. M. (1982). Functional groups bearing sulphur. In W. B. Jakoby, J. R. Bend and J. Caldwell (eds.), *Metabolic Basis of Detoxication*, Academic Press, New York, pp. 171–184.

Ziegler, D. M. (1984). Metabolic oxygenation of organic nitrogen and sulphur compounds. In J. R. Mitchell and M. G. Horning (eds.), *Drug Metabolism and Drug Toxicity*, Raven Press, New York, pp. 33–53.

Ziegler, D. M. (1988). Flavin-containing monooxygenase: catalytic mechanism and substrate specificities. *Drug Metab. Rev.*, **19,** 1–32.

7

Thioethers, thiols, dithioic acids and disulphides: Phase II reactions

John Caldwell and **Helen M. Given**
Department of Pharmacology and Toxicology, St. Mary's Hospital Medical
School, London W2 1PG, UK

SUMMARY

1. Although sulphur functional groups occur in a wide range of valency states and covalent bonding situations, normally only thiols and dithioic acids are involved in conjugations.
2. Thiols and dithioic acids can undergo metabolic conjugation with glucuronic acid, glucose and methyl groups, depending on the substrate and species in question; there is at present only tentative evidence for the formation of sulphate conjugates from sulphur (thiol) xenobiotics.
3. Disulphides are only substrates for conjugations after their initial reduction to thiols.
4. There are recent reports of xenobiotic thioethers affording sulphonium salts as metabolites; such compounds also exist in nature, and their biosynthesis in some cases does involve the sulphur lone pair.

7.1 CONJUGATION OF SULPHUR FUNCTIONAL GROUPS

Conjugation reactions involve the linkage of a xenobiotic to an endogenous moiety (known as the conjugating agent, or endocon) through a functional group present *per se* or introduced or revealed by a functionalization (phase I) reaction of oxidation, reduction or hydrolysis. The great majority of these conjugation reactions involve the replacement of a proton, present in a hydroxyl, amino or carboxyl group, by the conjugating agent. Rarely, the conjugating agent can form a covalent bond with a lone pair of electrons, as occurs in the metabolic quaternization of tertiary amines (NR′R″R‴) by methylation or glucuronidation (Caldwell, 1982a). Exceptions to the above occur in the case of glutathione conjugation, which can involve addition across suitably activated carbon–carbon double bonds, e.g. those present in α,β-unsaturated ketones, allylic esters etc., or nucleophilic displacement of halo- and nitro-alkanes and alkyl sulphonates (Caldwell, 1982b).

Sulphur functional groups occur in a very wide range of valency states and covalent bonding situations (see chapter 6, this volume). However, few of these functional groups are involved in conjugation reactions directly, and only two have suitably labile protons for replacement by an endogenous moiety, the thiol (–SH) and dithioic acid (–CSSH) groups. Although divalent sulphur in thioethers has two lone pairs of electrons, until recently these were thought not to be available for metabolic conjugation. Conjugation at the divalent sulphur would result in the production of sulphonium salts; if such metabolites are generated, their isolation and would be difficult in view of their chemical instability in some cases, and their polarity (see chapter 6, this volume, Part B). It is noteworthy that several sulphonium compounds exist in nature, and that their biosynthesis in some cases involves *S*-methylation at a divalent sulphur to produce the sulphonium species (see chapter 6, volume 1, Part B of this series). Sulphur lone pairs are more polarizable than those on oxygen or nitrogen and prefer to complex with soft Lewis acids (Hanzlik, 1984). The behaviour of the thiol group is considerably influenced by its empty d orbitals, which allow nucleophilic attack, notably by other thiols (Hanzlik, 1984). This is responsible for the thiol-disulphide interchanges which they commonly participate in, thus:

$$2R\text{–}SH \quad \rightleftharpoons \quad R\text{–}S\text{–}S\text{–}R$$
$$\text{thiol} \qquad\qquad \text{disulphide}$$

These redox reactions are facilitated by the fact that the outer electrons of sulphur are remote from the nucleus. Xenobiotics containing a disulphide linkage only undergo conjugations after an initial reduction to the thiol. In general terms, sulphur functionalities with low valencies tend to participate as nucleophiles in thiol–disulphide interchanges, while those with high valency states tend to be good leaving groups, being readily eliminated when displaced by nucleophiles, such as the thiol group of glutathione.

Thiols and dithioic acids are rarely present in xenobiotics as such and are not commonly produced by metabolism : alkyl thioethers, for example, undergo *S*-oxidation preferentially rather than dealkylation, which would give a free thiol. When thiols do occur, their availability for conjugation is limited by thiol–disulphide interchanges, giving dimers (homoconjugates) or mixed disulphides with cysteine, glutathione or protein sulphydryls (see chapter 4, Volume 2, Part B of this series).

There is little information in the literature on the conjugation reaction of thiols and dithioic acids. The metabolic behaviour of these groups has certain similarities to that of their oxygen analogues. They are known to undergo conjugation with glucuronic acid, which mechanism is replaced by glucosidation in molluscs, and methylation. They apparently do not undergo sulphation (see later) or acetylation, while dithioic acids can undergo amino acid conjugation in plant, but not animal, tissue. This chapter is a review of the available data on conjugation reactions of sulphur functionalities. From the comments above, it is clear that this has to be restricted to a description of conjugation reactions at thiol and dithioic acid functional groups.

7.2 *S*-GLUCURONIDATIONS

7.2.1 *S*-Glucuronidations of thiols

The first reported example of *S*-glucuronidation of a thiol group appears to be that of Parke (1952), who found thiophenol glucuronide as a metabolite of benzene in the rabbit. This was a very surprising finding at the time, but can now be rationalized in terms of the catabolism of the glutathione conjugate of benzene 1,2-oxide by the cysteine conjugates β-lyase pathway (see later and Volume 2, Part B, chapters 5 and 6 of this series). Following this, the first fully authenticated report of a xenobiotic *S*-glucuronide was that of Clapp (1956), arising from a study of the fate of benzothiazole-2-sulphonamide. This compound was synthesized during the course of the evaluation of a series of carbonic anhydrase inhibitors, and was very active in *in vitro* test systems. However, it had no diuretic activity when tested *in vivo*. This discrepancy was accounted for by the findings that benzothiazole-2-sulphonamide was rapidly and extensively metabolized by a novel pathway in which the 2-sulphonamido function was presumably transformed to a thiol function, leading ultimately to the elimination in urine benzothiazole-2-thiologlucuronide. Dogs given the parent sulphonamide excreted some 25% of the dose in the 0–24 h urine as this conjugate.

This unusual metabolic reaction was investigated further by Colucci and Buske (1965). These authors showed that this pathway was extant in rat and rabbit, and they confirmed its occurrence in the dog. The thioglucuronide was accompanied by two other compounds, benzothiazole-2-thiol and benzothiazole-2-mercapturate, as major urinary metabolites in all three species.

Colucci and Buyske (1965) also studied the mechanisms involved in the formation of these various metabolites and showed that the critical step is the displacement of the sulphonamide moiety by glutathione (see Scheme 3, chapter 7, volume 1, Part B of this series). The 2-glutathionyl benzothiazole so produced then undergoes successive hydrolyses to give 2-cysteinyl benzothiazole. This compound is then either *N*-acetylated, giving the mercapturic acid which is found in the urine, or cleaved at the C–S bond to give benzothiazole-2-thiol. This compound is excreted in the urine partly free and partly as its *S*-glucuronide. This study is apparently the first description of what is now realized to be a major catabolic route of glutathione conjugates, the cysteine conjugate β-lyase pathway (see Bakke, 1986, and Volume 2, Part B, chapter 5 and 6 this series). Following this, Illing and Benford (1976) showed that benzothiazole-2-thiol is a substrate for UDP-glucuronyl transferase(s) of rat liver microsomal preparations.

A great many thiopurines and thiopyrimidines have been tested as possible chemotherapeutic agents as nucleic acid base analogues, and a number of these have been shown to undergo metabolic *S*-glucuronidation. The first such findings concerned 9-ethyl- and 9-*n*-butyl-6-mercaptopurine, which are both conjugated with glucuronic acid through the thiol group (Hansen *et al.*, 1963).

The fate of 6-*n*-propyl-2-thiouracil (PTU), a widely used antithyroid compound, has been investigated in some detail. Marchant *et al.* (1971) were the first to show its conjugation with glucuronic acid, a β-glucuronidase-labile conjugate of the parent drug being found in the plasma and urine of rats and human subjects given PTU. Papapetrou *et al*. (1972) found a similar conjugate in the bile of rats given ^{35}S-PTU, accounting for 23% of biliary radioactivity, or 2% of the administered dose. Sitar and Thornhill (1972) found that 40–48% of a dose of PTU was excreted in the 0–24 h urine of rats in the form of a glucuronic acid conjugate of unchanged PTU. These authors also reported that the major biliary metabolite was also a glucuronide of PTU which had different characteristics from that found in the urine.

These findings were confirmed and extended by Lindsay *et al*. (1974), who had the advantage of working with both [35]S- and ^{14}C-labelled forms of the drug. This study confirmed that PTU glucuronide is the major metabolite in rat urine (42% of administered dose), and showed the presence of various other metabolites including *S*-methyl-PTU. In a further study, Lindsay *et al*. (1977) characterized the glucuronide of PTU found in rat urine and bile and that formed *in vitro* with guinea pig liver microsomal preparations. These studies have shown unequivocally that the glucuronic acid residue is attached to the PTU molecule through the thiol group in the 2-position, and that only one glucuronide is present. The chromatographic and certain other properties of the glucuronide were markedly influenced by the medium in which it was present, notably with the salt concentration: although bile is almost salt free, urine has a high salt content. These variations were suggested to be responsible for the earlier reports that more than one glucuronide of PTU is formed metabolically.

In recent years, a range of other *S*-glucuronides have been found as metabolites, mostly arising from the catabolism of glutathione conjugates of haloaromatic compounds (notably polychlorinated or polybrominated biphenyls). These are in general minor metabolites of the class of compounds (see Bakke, 1986).

The enzymology of glucuronidation of thiols has been examined by Dutton and Illing (1972) and Illing and Dutton (1973). Various features of the glucuronidation of thiophenols were compared with those of phenolic hydroxyl groups, using thiophenol, *o*-aminothiophenol and *p*-nitrothiophenol. It was found that the glucuronyl transferases behave in similar ways towards oxygen and and sulphur functionalities. Thus, the two activities have the same subcellular location, pH optima and exhibit latency, being activated by preincubation of microsomal preparations with detergents and by UDP-*N*-acetylglucosamine. The genetic defect in the glucuronidation of bilirubin, *o*-aminophenol and other substrates in the Gunn mutant of the Wistar rat is also manifest towards *o*-aminothiophenol, but not *p*-nitrothiophenol, paralleling the behaviour of their oxygen analogues. Additionally, like the corresponding phenol, glucuronidation of *o*-aminothiophenol by liver microsomal preparations from Gunn rats is restored by the addition of diethylnitrosamine. The induction of glucuronidation of thiophenols by phenobarbitone has been demonstrated in chick embryo *in ovo*.

Thioglucuronides have variable lability towards β-glucuronidase. Colluci and Buyske (1965) found that benzathioazole-2-thiolglucuronide was cleaved by bovine liver β-glucuronidase, while Dutton and Illing (1972) showed that the glucuronides of *p*-nitro- and *o*-amino-thiophenol were good substrates for the β-glucuronidase of rat preputial gland. However, these authors were unable to effect the hydrolysis of thiophenol glucuronide with β-glucuronidases from *E. coli*, rat liver or rat preputial gland. The reduced lability of certain thioglucuronides to β-glucuronidase was ascribed to the relatively greater length of C–S as compared with C–O bonds.

7.2.2 *S-Glucuronidation of dithioic acids*

Dithioic acids are the sulphur analogues of carboxylic acids and are only rarely encountered in xenobiotics. The only such compound to be examined systematically in mammals is diethyldithiocarbamic acid (DDC), the active metabolite of the alcohol dehydrogenase inhibitor disulfiram. During an early study on the glucuronidation of methanol and ethanol in rabbits, which involved the administration of disulfiram to inhibt the oxidative breakdown of these compounds, Kamil *et al.* (1953) reported evidence for the excretion of small amounts of an ester-type glucuronide of DDC. Some 10 years later, Kaslander (1963) showed that this glucuronide was a minor urinary metabolite of disulfiram in humans, and Stromme (1965) found this glucuronide in rat urine. In the rat, concurrent administration of ethanol reduced the rate of glucuronidation. In a later study, Gessner and Jakubowski (1972) showed that DDC *S*-glucuronide was the major urinary metabolite of both DDC and disulfiram in the rat, accounting for some 30% of the administered dose. A full account of disulfiram metabolism is given in chapter 8, volume 3, Part B of this series.

Dutton and Illing (1972) have examined the enzymology of this thioester glucuronidation and have shown that its mechanism is essentially identical with that operating for oxygen functional groups. The thioester glucuronide of DDC is readily hydrolysed by rat preputial gland β-glucuronidase, and this activity is inhibited by saccharo-1,4-lactone.

7.3 *S*-METHYLATIONS

7.3.1 *S-Methylations of thiol functions*

The *S*-methylation of thiol groups is now recognized to be an important pathway in their biotransformation. As has been mentioned already, thiols are but rarely encountered *per se* in xenobiotics, but this conjugation is of major significance in the so-called 'thiomethyl shunt' in the catabolism of cysteine conjugates (Jakoby and Stevens, 1984, and see chapters 5 and 6 in Volume 2, Part B of this series).

The first reported example of metabolic methylation of a thiol group appears to be that of thiouracil (Sarcone and Sokal, 1958). The bulk (>80% of administered dose) of this compound is eliminated unchanged by rats, but some 8% is recovered in the form of the *S*-methyl conjugate. Further examples of this type of *S*-methylation occur in the metabolism of *S*-methyl-6-*n*-propyl-2-thiouracil in the rat (Lindsay *et al.*, 1974), captopril in human and rat tissues (Drummer *et al.*, 1983) and D-penicillamine in man (Perrett *et al.*, 1976). In each case, this metabolic reaction appears to be a very minor one, and the urinary excretion of the *S*-methyl conjugate accounts for less than 5% of the dose.

In the course of studies on choline biosynthesis, Bremer and Greenberg (1961) discovered that a variety of simple, low molecular weight thiol compounds were transmethylated by an enzyme system present in hepatic microsomes of rat and six other mammalian species. These xenobiotic thiols included mercaptoethanol, mercaptoacetic acid, β-mercaptopropionic acid, methyl mercaptan and hydrogen sulphide, and in each case the methyl group was transferred from S-adenosylmethionine. Endogenous thiol compounds such as glutathione, cysteine and homocysteine were not methylated, while the replacement of S-adenosylmethionine with S-adenosylethionine resulted in transethylation. This enzyme system has been studied in more detail recently in relation to the fate of thiols produced by cysteine conjugate β-lyase and is discussed extensively in Volume 2 of this series (chapters 6 and 7).

The enzymic basis of the S-methylation of thiopurines and thiopyrimidines was first examined by Remy (1963). He showed that the methyl group transferred to the thiol was derived from S-adenosylmethionine and found that a considerable number of 2-, 6- and 8-thiopurines and 2-, 4- and 5-thiopyrimidines acted as a acceptor substrates for thiomethyltransferases present in rat and mouse liver and kidney. More recently, Woodson and Weinshilboum (1983) have purified a thiopurine methyltransferase from the cytosol of human kidney, with the molecular weight of about 36 kD. This enzyme methylates a range of thiopurines and thiopyrimidines, but aliphatic thiol compounds are very poor, or completely inactive, as methyl acceptors. This thiopurine methyltransferase was clearly differentiated from the thiol methyltransferase of human kidney on the basis of its subcellular distribution, substrate specificity, kinetic characteristics and differential sensitivity to inhibitors.

7.3.2 S-Methylation of dithioic acids

Gessner and Jakubowski (1972) have reported an unusual pathway of metabolism of disulfiram in rats, leading to the formation of the thiomethyl ester of DDC. DDC-methyl ester was found in rat urine after administration of disulfiram, while microsomal preparations from rat liver and kidney contained an active methyltransferase, using S-adenosylmethionine as methyl donor.

Oxidation to sulphate ion is a major metabolic pathway of disulfiram and DDC, and studies *in vivo* have shown that DDC-methyl ester is extensively (>60% of dose) metabolized along this pathway. In contrast, administration of DDC itself gave rise to only 16% of dose as inorganic sulphate. It thus seems possible that the S-methylation of this dithioic acid is an intermediate in its oxidation to sulphate.

7.3.3 S-Methylation of thioethers

A novel pathway of metabolism, identified recently by Hoffman *et al.* (1988) and Mozier *et al.* (1988), is where the lone pair of electrons on divalent sulphur in thioethers participates in metabolic conjunction with methyl groups to afford methylsulphonium ions as metabolites. The cytosolic enzyme mediating this reaction has been named 'thioether S-methyltransferase' or 'TEMTase', and it utilizes S-adenosylmethionine as a methyl donor. TEMTase from mouse lung has been purified, and shown to be a monomer of $M_\gamma \approx 28\,000$ with a pI of 5.3, a pH optimum of 6.3, and with broad substrate specificity for various thioethers. Activity measurements and immunoblotting techniques indicate that TEMTase is present in all mammalian species tested to date. This would indicate that TEMTase may play an important role in the metabolism of thioethers to more water-soluble methyl sulphonium ions for excretion in urine.

7.4 *S*-GLUCOSIDATIONS

It is well known that the glucuronic acid conjugation mechanism is replaced in insects and other invertebrates by the analogous reaction of glucoside formation (Smith, 1977). In this reaction, the glucose used is derived from UDP-glucose and the transferase effecting the reaction is found in the microsomal fraction of the cell. Dutton and Illing (1972) have shown that intact gut cells and homogenates from the slug *Arion ater* carry out the *S*-glucosidation of thiophenol and of DDC, in ways essentially identical with the *O*-glucosidation of analogous substrates. The UDP-glucosyltransferase of *Arion ater* is inducible by phenobarbitone (Leakey and Dutton, 1975). Neither of these thioglucosides was a substrate for sweet almond emulsin (β-glucosidase) and they also exhibited a marked resistance to acid hydrolysis (Dutton and Illing, 1972).

The glucosidation of thiols by insects has been studied by Gessner and Acara (1968). The *S*-glucosidation of thiophenol and of 5-theoracil was found to occur *in vivo* in two species of crickets and two of cockroaches, and their *S*-glucosides were found in the excreta. Studies *in vitro* have shown that this conjugation is effected by enzymes present in the fat bodies of the insects in question and uses UDP-glucose as glucose donor. As might be expected, glucose itself was unable to function as a precursor to the conjugate. In addition, Gessner and Acara (1968) carried out experiments to compare the **O-** and *S*-glucosidation of these substrates and found the two pathways had comparable turnover rates and, of course, used the same cofactor, i.e. UDP-glucose.

7.5 THIOAMIDE FORMATION

As has been mentioned above, the dithioic acid group is the sulphur analogue of the carboxyl moiety. The major metabolic reactions of carboxylic acids are glucuronidation and amino acid conjugation (Caldwell *et al.*, 1987). Although the occurrence of thioester glucuronidation is well established in mammals, at least in the case of DDC, there appear to be no reports of the amino acid conjugation of dithioic acids in animal species. However, it is of interest to note that dimethyldithiocarbamic acid is converted to a small extent to the corresponding alanine thioamide conjugate by slices of potato tuber (Kaslander *et al.*, 1962). This amino acid conjugation occurs in addition to the formation of the thioester glucoside.

7.6 SULPHATE CONJUGATIONS

The drug metabolism literature has long alluded to the conversion of thiophenols to thiosulphate conjugates, on the basis of an early statement to the effect that the administration of thiophenol to rabbits led to an increase in the urinary excretion of 'ethereal sulphates' (see Williams, 1959). However, it is important to appreciate that this statement was until recently entirely unsubstantiated and that there was no reported evidence for the conversion of any xenobiotic thiol to such thiosulphate conjugates (Mulder, 1980). In very recent studies, Miwa *et al.* (1988) have provided some tentative evidence for the enzymatic *S*-sulphate conjugation of xenobiotic thiols. They demonstrated that [^{35}S]4-nitrobenzyl mercaptan (NBM) was enzymically transformed into its *S*-sulphate at pH 6.0 in rat liver cytosol fortified with PAPS under anaerobic conditions in the presence of EDTA. Further studies are required to ascertain the role of *S*-sulphation in the metabolism and excretion of xenobiotic thiols.

Thiolsulphates of endogenous compounds play essential roles in intermediary metabolism, most notably by their participation in the sulphate sulphur pool and its reactions. These matters are described in detail elsewhere in this series (see chapter 5 of volume 1, Part A and chapter 3 of Volume 2, Part B).

7.7 CLOSING COMMENTS

This short review has shown that the thiol and dithioic acid groups can undergo metabolic conjugation with glucuronic acid, glucose and methyl groups, depending on the structure and animal species in question. This information has been obtained sporadically over a period of some 30 years and, although the considerable attention paid in recent years to the catabolism of glutathione conjugates has added substantially to our knowledge, there remains an outstanding need for systematic studies to define more precisely the chemical and biological factors influencing such *S*-conjugations.

REFERENCES

Bakke, J. E. (1986). Catabolism of glutathion conjugates. In G. D. Paulson, J. Caldwell, D. H. Hutson and J. J. Menn (eds.), *Xenobiotic Conjugation Chemistry*, ACS Symposium Series 299, Washington, DC, pp. 301–321.

Bremer, J. and Greenberg, D. M. (1961). Enzymic methylation of foreign sulfhydryl compounds. *Biochim. Biophys. Acta, 46*, 217–224.

Caldwell, J. (1982a). Conjugation reactions of nitrogen centres. In W. B. Jakoby, J. R., Bend and J. Caldwell (eds.), *Metabolic Basis of Detoxication*, Academic press, New York, pp. 291–306.

Caldwell. J. (1982b). Conjugation reactions in the metabolism of xenobiotics. In I. M. Arias, H. Popper, D. Schachter and D. A. Shafritz (eds.) *The Liver: Biology and Pathobiology*, Raven Press, New York, pp. 281–295.

Caldwell, J., Weil, A. and Sinclair, K. A. (1987). Acylation of amino acids and other endobiotics by xenobiotic carboxylic acids. In J. W. Gorrod, H. Oehlschlager and J. Caldwell (eds.), *Metabolism of Xenobiotics*, Taylor and Francis, London, pp. 217–224.

Clapp, J. W. (1956). A new metabolic pathway for a sulfonamide group. *J. Biol. Chem., 223*, 207–214.

Colucci, D. F. and Buyske, D. A. (1965). The biotransformation of a sulfonamide to a mercaptan and to mercapturic acid and glucuronide conjugates. *Biochem. Pharmacol., 14*, 457–4366.

Dutton, G. J. and Illing, H. P. A. (1972). Mechanism of biosynthesis of thio-beta-D-glucuronides and thio-beta-D-glucosides. *Biochem. J., 129*, 5439–550.

Drummer, O. H., Miach, P. and Jarrott, B. (1983). *S*-methylation of captopril. Demonstration of captopril thiol methyltransferase activity in human erythrocytes and enzyme distribution in rat tissues. *Biochem. Pharmacol.*, **32**, 1557–1562.

Gessner, T. and Acara, M. (1968). Metabolism of thiols: beta-glucosidation. *J. Biol. Chem.*, **243**, 3142–3147.

Gessner, T. and Jakubowski, M. (1972). Diethyldithiocarbamic acid methyl ester. A metabolite of disulfiram. *Biochem. Pharmacol.*, **21**, 219–230.

Hansen, H. J., Giles, W. G. and Nadler, S. B. (1963). Metabolism of 9-ethyl-6-MP-S[35] and 9-butyl-6-MP-S[35] in humans. *Proc. Soc. Exp. Biol. Med.*, **113**, 163–165.

Hanzlik, R. (1984). Prediction of metabolic pathways – sulfur functional groups. In J. Caldwell and G. D. Paulson (eds.) *Foreign Compound Metabolism*, Taylor and Francis, London, pp. 65–78.

Hoffman, J. L., Mozier, N. M. and Warner, D. R. (1988). *S*-Adenosylmethionine:thioether *S*-methyltransferase (TEMTase) in the metabolism of sulphur xenobiotics. *Abstracts of the 1st International Symposium on Sulphur Xenobiochemistry, London*, abstract 023.

Illing, H. P. A. and Dutton, G. J. (1973). Some properties of the uridine diphosphate glucuronyltransferase activity synthesizing thio-*beta*-D-glucuronides. *Biochem. J.*, **131**, 139–147.

Illing, H. P. A. and Benford, D. J. (1976). Observations on the accessibility of acceptor substrates to the active centre of UDP-glucuronosyltransferase. *Biochim. Biophys. Acta*, **429**, 768–779.

Jakoby, W. B. and Stevens, J. (1984). Cysteine conjugate beta-lyase and the thiomethyl shunt. *Biochem. Soc. Trans.*, **12**, 33–35.

Kamil, I. A., Smith, J. N. and Williams, R. T. (1953). Studies in detoxication. 50. The isolation of methyl and ethyl glucuronides from the urine of rabbits receiving methanol and ethanol. *Biochem J.*, **54**, 390.

Kaslander, J., Kaars Sijpesteijn, A. and Van der Kerk, G. J. M. (1962). On the transformation of the fungicide sodium dimethyldithiocarbamate into its alanine derivative by plant tissues. *Biochim. Biophys. Acta* **60**, 417–419.

Kaslander, J. (1963). Formation of an S-glucuronide from tetraethylthiuram disulfide (Antabuse) in man. *Biochim. Biophys. Acta* **71**, 730–732.

Leakey, J. E. A. and Dutton, G. J. (1975). Effect of phenobarbital on UDP-glucosyltransferase activity and phenolic glucosidation in the mollusc *Arion ater*. *Comp. Biochem. Physiol.*, **51C**, 215.

Lindsay, R. H., Hill, J. B., Kelly, K. and Vaughn, A. (1974). Excretion of propylthiouracil and its metabolites in rat bile and urine. *Endocrinology*, **94**, 1689–1698.

Lindsay, R. H., Vaughn, A., Kelly, K. and Abou-Enein, H. Y. (1977). Site of glucuronide conjugation to the antithyroid drug 6-*n*-propyl-2-thiouracil. *Biochem. Pharmacol.*, **26**, 833–840.

Marchant, B., Alexander, W. D., Robertson, J. W. K. and Lazarus, J. H. (1971). Concentration of ^{35}S-propylthiouracil by the thyroid gland and its relationship to anion trapping mechanism. *Metabolism*, **20**, 989–999.

Miwa, K., Okuda, H. and Watabe, T. (1988). The *S*-sulphate formation from 4-nitrobenzyl mercaptan by rat liver cytosolic sulphotransferase and its covalent binding to the sulphydryl-groups of the cytosolic proteins. *Abstracts of the IInd International ISSX Meeting, Kobe*, Abstract II. 404–P7.

Mozier, N. M., McConnell, K. P. and Hoffman, J. L. (1988). S-Adenosyl-L-methionine: Thioether S-Methyltransferase, a new enzyme in sulphur and selenium metabolism. *J. Biol. Chem.*, **263**, 4527–4531.

Mulder, G. J. (1980). In G. J. Mulder (ed.) *Sulfation of Drugs and Related Compounds*, CRC Press, Boca Raton, FL, p. 19.

Papapetrou, P. D., Marchant, B., Gavras, H. and Alexander, W. D. (1972). Biliary excretion of ^{35}S-labelled propylthiouracil, methimazole and acarimazole in untreated and pentobarbitone pretreated rats. *Biochem. Pharmacol.*, **21**, 363–377.

Park, D. V. (1952). The metabolism of aromatic compounds. *Ph.D. Thesis*, University of London.

Perrett, D., Snedden, W. and Stephens, A. D. (1976). Studies on D-penicillamine metabolism in cystinuria and rheumatoid arthritis: Isolation of *S*-methyl-D-penicillamine. *Biochem. Pharmacol.* **25**, 259–264.

Remy, C. N. (1963). Metabolism of thiopyrimidines and thiopurines. S-methylation with S-adenosylmethionine transmethylase and catabolism in mammalian tissues. *J. Biol. Chem.*, **238**, 1078–1084.

Sarcone, E. J. and Sokal, J. E. (1958). Detoxication of thiouracil by S-methylation. *J. Biol. Chem.*, **231**, 605–608.

Sitar, D. S. and Thornhill, D. P. (1972). Propylthiouracil: absorption, metabolism and excretion in the albino rat. *J. Pharmacol. Exp. Ther.*, **1834**, 440–448.

Smith, J. N. (1977). Comparative detoxication of invertebrates. In D. V. Parke and R. L. Smith (eds), *Drug Metabolism from Microbe to Man*, Taylor and Francis, London, pp. 219–232.

Stromme, J. H. (1965). Metabolism of disulfiram and diethyldithiocarbamate in rats with demonstration of an *in vivo* ethanol-induced inhibition of the glucuronic acid conjugation of the thiol. *Biochem. Pharmacol.*, **14**, 393–410.

Williams, R. T. (1959). *Detoxication Mechanisms*, 2nd ed., Chapman and Hall, London, p. 492.

Woodson, L. C. and Weinshilboum, R. M. (1983). Human kidney thiopurine methyltransferase. Purification and biochemical properties. *Biochem. Pharmacol.*, **32**, 819–826.

Index